AS/A-Level

Physics

Mike Crundell

Philip Allan Updates
Market Place
Deddington
Oxfordshire
OX15 0SE

tel: 01869 338652
fax: 01869 337590
e-mail: sales@philipallan.co.uk
www.philipallan.co.uk

© Philip Allan Updates 2001

ISBN 0 86003 775 4

Cover illustration by John Spencer

Printed by Raithby, Lawrence & Co Ltd, Leicester

Contents

Introduction

■ ■ ■

Questions and answers

Introduction

About this guide

This book is intended to give you experience of what it is like to answer examination questions. It is not just a book of examination questions, because answers are also provided and these are assessed from an examiner's viewpoint.

Each question is followed by answers by two candidates, A and B. Candidate A is an A-grade student whereas candidate B can expect to be awarded a much lower grade. The answers are accompanied by comments from the examiner. These are preceded by the icon [e] and indicate what has gone wrong, where marks have been lost and where marks could have been gained. General hints for the improvement of answers are also provided.

The book is divided into a number of sections. Sections 1–3 deal with subject material that is found in the AS section of the UK specification. Sections 4–11 contain material that is found in either AS or A2, depending on which specification you are following. The remainder of the sections (12–18) are relevant to the A2 part of the specification.

The values of constants are given with each question in which they are required. You should remember that most examination papers will have a separate sheet providing all the data. Also, there may be another sheet with some formulae. Look at some specimen papers so that you know which data and formulae are provided. To save time in the exam, find out where to find the information on the various pages.

Although the answers are provided in this book, you should be able to gain valuable experience by doing the questions for yourself and then comparing your responses with those of candidates A and B.

Question structure

Very few questions on examination papers are, in fact, questions! You will notice that, in many cases, each section of the 'question' begins with a *command word* (see page 2) and then ends by indicating how many marks have been allocated to the section.

The command word makes it clear what the examiners require. The mark at the end of each part of the question gives you a good idea of how much time you should spend on it. Notice how many marks there are for the whole examination and the length of time allowed. Divide the total mark by the time and you have an idea of how long to spend for each mark. Many examinations work out at about one mark per minute.

Command words

Define...

Give a formal statement or the defining equation, with symbols identified. No further comment is required.

Explain.../What is meant by...?

In general, the definition should be given, together with some relevant comment. The amount of comment required depends on the number of marks allocated.

List...

A number of particular points should be made, with no supporting comments. Frequently, the number of points to be made in the list will be stated. Do not exceed that number.

State...

The answer should be given with very little or no supporting comment.

Sketch...

This can be applied to a graph or a diagram. On a sketch graph, the axes should be labelled. Important features, such as the origin, intercept and curvature of the line (if any), are required. Sketch diagrams should show important features and should indicate proportions.

Calculate...

A numerical answer is required. It is important to show working and to give the unit of the answer. If the answer only is required, then other commands, such as *state*, would be used.

Determine...

This is used in the same way as *calculate*. The quantity to be determined cannot be found directly but may be obtained by substitution into a formula, for example.

Estimate...

This command is used if the quantity cannot be determined precisely. Candidates are expected to make reasonable assumptions and give sensible values of quantities not listed in the question.

Describe...

Candidates are expected to state the features of the subject matter. This command word is often used in the context of experiments or phenomena. Diagrams should be drawn, where appropriate. The length of any description should be assessed by reference to the mark allocation. In general, descriptions should be in continuous prose, not in note form. Marks are awarded in the examination for 'quality of written communication'.

Outline...

Used in a similar context to *describe* but generally implying that less detail is required.

Show...

The answer may be based on a mathematical deduction or on the calculation of a stated quantity. It is very important that terms are clearly defined and procedures are explained.

Discuss...

Candidates are expected to give a reasoned argument based on information relevant to the topic.

Suggest...

This may imply that there is no unique answer or that knowledge and understanding gained from within the specification are to be used to solve a novel situation. In general, a reasoned argument is required but the extent of the argument will be governed by the mark allocation.

Some final advice

Examiners all too often find that candidates have lost marks needlessly. Some of the following advice may seem to be very elementary, but it is important.

- **Read the question carefully.** Take special care to note the command words and the mark allocations.
- **Units.** The correct unit must be given with any numerical answer.
- **Meaning.** Always look at the answer to a calculation and decide whether it is sensible — it is all too easy to make a power-of-ten error.
- **Significant figures.** The number of significant figures in an answer should be the same as that of the data. In the theory papers, data are usually given to either two or three significant figures. Examiners usually allow an answer to be given with one significant figure more than is appropriate for the given data.
- **Explanation.** Time in the examination is limited, but work should be explained. In any calculations, write down the equations being used before carrying out any algebraic manipulation or substitution. Remember that 'a diagram speaks a thousand words'. Wherever appropriate, sketch a diagram but remember that it must not be so rough as to be worthless!

Statics

Question 1.1

(a) Distinguish between a *scalar* and a *vector* quantity. (2 marks)

(b) State the conditions necessary for an object to be in equilibrium. (3 marks)

(c) A square uniform trapdoor has sides of length 120 cm and has a weight of 45 N. It is hinged on a wall and is held open at an angle of 40° to the horizontal by means of a rope attached as shown in the diagram below.

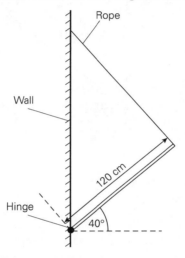

The rope is at right angles to the trapdoor.
(i) Copy the diagram and mark, with an arrow labelled *W*, the weight of the trapdoor. Also, mark and label with the letter *F* the force on the trapdoor due to the hinge.
(ii) Determine the tension in the rope. (5 marks)

Total: 10 marks

■ ■ ■

Answer to Question 1.1: candidate A

(a) A scalar has magnitude only. A vector has magnitude and direction.

🄔 2 marks. Magnitude and direction have been mentioned.

(b) The algebraic sum of the forces in any direction is zero. The clockwise moments are equal to the anticlockwise moments about any point.

🄔 3 marks. It would have been better to talk about the sum of the moments about any point.

(c) (i)

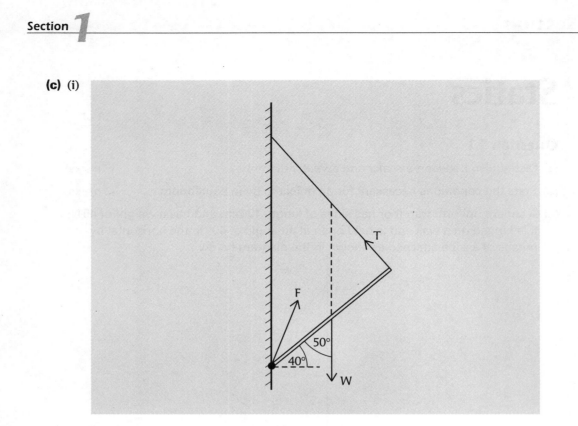

(ii) Taking moments about the hinge

45sin50 × 60 = T × 120

T = 17.2 N

e 5 marks. In part (i) *W* is shown correctly. *F* points towards the point of intersection of *W* and the tension in the rope. (Remember that, for three co-planar forces to be in equilibrium, they must pass through one point.) The calculation in part (ii) is correct. Note that the candidate marked all forces and angles on the diagram before starting the calculation.

■ ■ ■

Answer to Question 1.1: candidate B

(a) Only a vector has direction.

e 1 mark only. The examiners would expect some mention of the other aspect, i.e. magnitude.

(b) Forces acting up equal forces acting down, forces to right equal forces to left. Sum of clockwise moments equals sum of anticlockwise moments.

e 2 marks. The first statement is clumsy, but essentially correct. No mention has been made of the fact that moments may be taken about any point.

(c) (i)

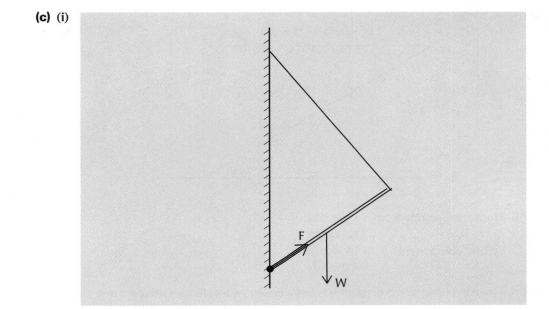

(ii) $W\sin40 \times 60 = T \times 120$
$T = 14.5\,\text{N}$

e 2 marks. W has been shown acting at the centre (approximately!) but the direction of F is incorrect. The candidate had the idea of taking moments, but the component of W normal to the trapdoor has not been used. Always draw a sketch diagram with all angles and distances marked.

Question 1.2

(a) State which of the following are vector quantities: distance; acceleration; mass; energy; force; temperature. (2 marks)

(b) A ball is thrown into the air. When it lands, its speed is $14\,\text{m s}^{-1}$ at an angle of $30°$ to the horizontal.
  (i) Draw a scale diagram to determine the horizontal component and the vertical component of the velocity of the ball at the time of landing.
  (ii) Check your answers in (i) by calculation. (5 marks)

Total: 7 marks

■ ■ ■

Answer to Question 1.2: candidate A

(a) The vectors are acceleration and force.

e 2 marks. Both vectors correctly identified.

(b) (i)

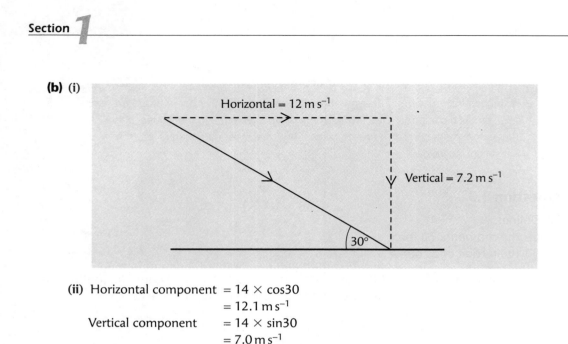

(ii) Horizontal component $= 14 \times \cos30$
$= 12.1\,\mathrm{m\,s^{-1}}$
Vertical component $= 14 \times \sin30$
$= 7.0\,\mathrm{m\,s^{-1}}$

e 4 marks. In part (i) the diagram has been drawn with sufficient accuracy and the answers are within tolerance, but the scale has not been given. The calculations in part (ii) are correct.

▪ ▪ ▪

Answer to Question 1.2: candidate B

(a) Acceleration, energy, force.

e 1 mark. Although both vectors have been identified, a scalar (energy) has also been included.

(b) (i)

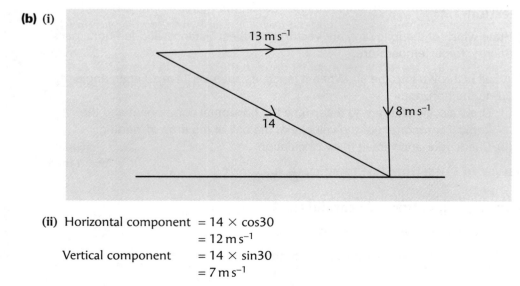

(ii) Horizontal component $= 14 \times \cos30$
$= 12\,\mathrm{m\,s^{-1}}$
Vertical component $= 14 \times \sin30$
$= 7\,\mathrm{m\,s^{-1}}$

e 3 marks. The basic shape of the vector diagram is correct but the angles and distances have not been drawn with sufficient care. Always use a protractor and ruler for such diagrams. One further mark could have been lost because, although the answers are correct, a one significant figure answer has been given from two significant figure data.

Question 1.3

(a) Define:
  (i) the moment of a force
  (ii) the torque of a couple (4 marks)

(b) Two forces, each of magnitude 15 N, act in opposite directions at right angles to a rod of length 60 cm, as illustrated in the diagram below.

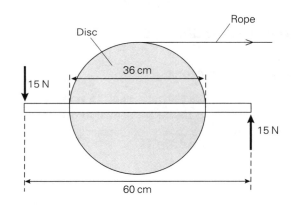

A disc of diameter 36 cm is fixed to the rod so that their centres coincide. A rope, attached to the disc and held tangential to it, prevents the disc turning. Calculate the tension in the rope. (4 marks)

(c) Suggest, with a reason, what must be done to the tension in the rope in order to prevent the disc turning if the rope is no longer held so that it is tangential to the disc. (3 marks)

Total: 11 marks

■ ■ ■

Answer to Question 1.3: candidate A

(a) (i) Force × perpendicular distance from force to pivot.
  (ii) Force × perpendicular distance between the two forces.

  e 4 marks. In part (i) it would have been better to refer to the perpendicular distance between the line of action of the force and the pivot. In part (ii) there is a mention that two forces are involved.

(b) Torque of couple = 15 × 0.60 = 9.0 N m
    9 = tension × 18
    $T = 0.5$ N

> e 3 marks. The torque of the couple has been equated with the moment of the tension but the candidate failed to convert centimetres to metres when working out the moment. The correct answer is 50 N.

(c) The tension must be increased because the distance of the tension from the centre of the disc will decrease. If distance decreases, force must increase to keep force × distance constant.

> e 3 marks. An acceptable answer. It would have been better to refer to the *line of action* of the tension.

■ ■ ■

Answer to Question 1.3: candidate B

(a) (i) Force × distance.
    (ii) Force × distance between the forces.

> e Only 1 mark. The definition in part (i) could apply to work done! It is important to stress that the force and the distance are normal to one another. Also, distance from what? In part (ii), once again, perpendicular distance has not been mentioned.

(b) Torque = couple
    15 × 60 = T × 36
    $T = 25$ N

> e 3 marks. The distances have not been converted to metres, but this is not necessary because all distances are in centimetres. The diameter of the disc has been used, rather than the radius, when calculating the moment of the tension.

(c) The distance from the pivot increases and so the tension must be changed to keep force × distance constant.

> e 1 mark only. The candidate realises that the moment of the tension must remain constant. However, the distance is said to increase and this is incorrect. Furthermore, the tension is said to 'change'. Stating 'change' will rarely be given any marks. The direction of any change must be made clear.

Dynamics

Question 2.1

(a) Define acceleration. (2 marks)

(b) The diagram below shows the variation with time t of the speed v of a sphere as it falls vertically.

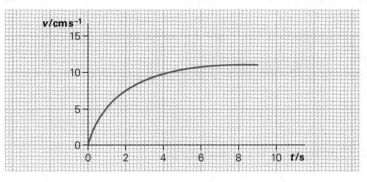

Use the diagram to determine the acceleration:
(i) at time $t = 2.0\,s$
(ii) at time $t = 8.0\,s$ (4 marks)

(c) Discuss the evidence provided in the diagram for the statement that the sphere is falling in a resistive fluid. (4 marks)

Total: 10 marks

Answer to Question 2.1: candidate A

(a) Acceleration is the change in velocity per unit time.

> 2 marks. Reference has been made to velocity, not speed, and the ratio is clear. However, it is often wiser to write out the relation as a word equation.

(b) (i)

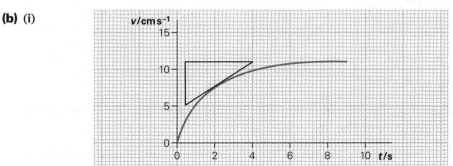

$$\text{Gradient} = \text{acceleration} = \frac{11.0 - 5.0}{4.0 - 0.4}$$

$$\text{acceleration} = 1.7 \, \text{cm} \, \text{s}^{-2}$$

(ii) Zero.

 3 marks. Although the tangent to the curve is acceptable, the triangle used to determine the gradient is too small and consequently the result lacks accuracy. The answer in (ii) can be obtained by inspection.

(c) In a resistive fluid, resistance to motion increases with speed. Therefore, the accelerating force decreases as speed increases. The speed increases to a steady value. The graph shows decreasing acceleration to give a steady speed.

 4 marks. A good answer with all the points covered. Note that a common mistake is to think that, since the acceleration is decreasing, the speed must be decreasing also!

■ ■ ■

Answer to Question 2.1: candidate B

(a) The change in speed in one second.

 No marks. Acceleration is defined in terms of change of velocity, not speed. The ratio is not clear. Furthermore, the candidate has attempted the definition in terms of units, not quantities. Units should not be used when defining a quantity.

(b) (i) $\text{Acceleration} = \dfrac{7.5}{2.0}$

$= 3.75 \, \text{cm} \, \text{s}^{-2}$

(ii) $0 \, \text{cm} \, \text{s}^{-2}$

 1 mark only. In part (i) the gradient has not been used. Instead, and quite wrongly, the coordinates of a point on the line have been used. The mark has been given for part (ii).

(c) The acceleration decreases as time goes by and the sphere reaches terminal speed. It must be a resistive fluid.

 2 marks. The answer is a correct interpretation of the graph. However, there is no explanation as to how the graphical interpretation leads to the required conclusion.

Question 2.2

(a) Explain how it is possible for a body to accelerate in a direction which is not the same as that of its velocity. (4 marks)

(b) Outline an experiment to determine the acceleration of free fall, using a free-fall method. (6 marks)

(c) A ball is falling in a vacuum on the surface of the Moon. In a time of 1.50 s, it falls 180 cm from rest. Calculate:
  (i) the acceleration of free fall at the Moon's surface
  (ii) the time taken for the ball to fall a further 20 cm

  (5 marks)

  Total: 15 marks

Answer to Question 2.2: candidate A

(a) In circular motion, the velocity is along the tangent to the circle but the acceleration is towards the centre of the circle. This is because the direction is changing.

> 2 marks. This is a good explanation. However, the candidate has not included the situation in which the object is slowing down.

(b)

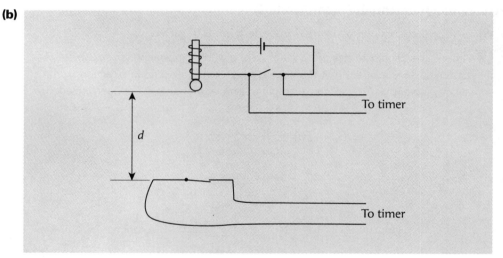

To timer

To timer

The distance between the ball on the magnet and the trapdoor is measured with a metre ruler. When the ball is released, the timer starts. The timer stops when the ball opens the trapdoor. The distance and the time are noted. The experiment is repeated for different values of d and t. Since $d = \frac{1}{2}gt^2$, plotting a graph of d against t^2, the gradient is $\frac{1}{2}g$.

> 6 marks. This is a good outline. It is clear how the measurements are made and how they are processed to find g. It would have been better if a sketch graph had been included so that there could be no doubt as to what quantity is plotted on each axis. Note that marks would have been lost if the theory had been omitted or the determination had been based on just one set of measurements.

(c) (i) $s = ut + \frac{1}{2}at^2$
  $1.8 = \frac{1}{2}g \times 1.5^2$
  $g = 1.6\,\text{m s}^{-2}$

(ii) Time to fall 2.0 m

$2.0 = \frac{1}{2} \times 1.6 \times t^2$

$t = 2.5\,s$

Time to fall further 20 cm = 2.5 − 1.5 = 1.0 s

e 4 marks. Part (i) is correct. In part (ii) the candidate has made a very common error — the value shown for t is in fact t^2. The answer should be $\sqrt{2.5 - 1.5} = 0.08\,s$. The problem could have been solved using the equation $v = u + at$ and finding $v = 2.4\,m\,s^{-1}$ when s = 1.8 m. Then, by substitution into $s = ut + \frac{1}{2}at^2$, $0.2 = 2.4t + \frac{1}{2} \times 1.6 \times t^2$. However, this does involve the solution of a quadratic equation. If you are not familiar with this technique, do not use the method.

■ ■ ■

Answer to Question 2.2: candidate B

(a) When the object is slowing down, the acceleration is negative.

e 1 mark. The candidate has the correct idea, but the explanation is incomplete. A statement to the effect that the acceleration is in the opposite direction to the velocity is necessary. Also, the example of circular motion has not been included.

(b)

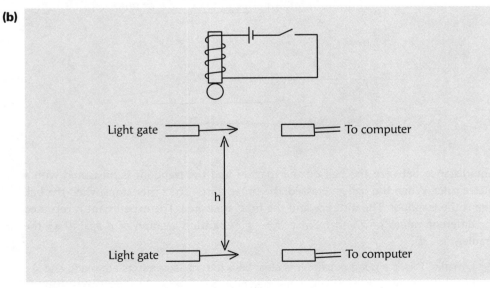

The light gates are connected to the computer and the distance between the gates is measured and fed into the computer. When the ball falls, the computer works out the speed as it falls through each gate. The computer then gives the acceleration.

e 1 mark. This answer is typical of many where IT packages have been used. It is important to realise that the candidate must explain to the examiner what readings have to be taken and how they are processed. Be warned!

(c) (i) $s = ut + \frac{1}{2}at^2$

$1.8 = \frac{1}{2}g \times 1.5^2$

$g = 1.6\,\text{m s}^{-2}$

(ii) To fall a further $20\,\text{cm}$

$0.2 = \frac{1}{2} \times 1.6 \times t^2$

$t^2 = 0.25$

$t = 0.5\,\text{s}$

🄴 2 marks. Part (i) is correct. A common mistake has been made in part (ii). The candidate has assumed that, when travelling the further $0.2\,\text{m}$, the ball started from rest.

Question 2.3

(a) One of the equations of motion may be expressed in the form

$$s = ut + \frac{1}{2}at^2$$

(i) Explain the meaning of each symbol in the equation.

(ii) State the conditions necessary for this equation to apply. (4 marks)

(b) A student carries out an experiment involving the free fall of a steel sphere, as illustrated below.

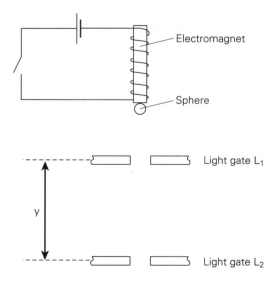

When the sphere is released from the electromagnet, it falls between two light gates L_1 and L_2. The light gate L_2 may be moved vertically. The time t to travel between L_1 and L_2 is noted, together with their separation y. The student then plotted the graph shown.

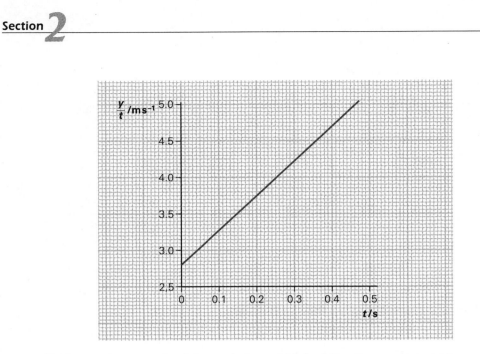

(i) State why a graph of the variation with t^2 of the distance y would not give a straight line.

(ii) Show that

$$\frac{y}{t} = u + \tfrac{1}{2}at$$

(iii) Use the graph and the result in (ii) to determine:
   (1) the speed of the sphere as it passes through light gate L_1
   (2) the acceleration of free fall
   (3) the distance the sphere falls before reaching L_1 (9 marks)

Total: 13 marks

■ ■ ■

Answer to Question 2.3: candidate A

(a) (i) u is the initial speed and s is the distance moved in time t with an acceleration a.
   (ii) The acceleration is constant.

   🄔 3 marks. Part (i) is correct, but in part (ii), the candidate has not mentioned that the equation applies to motion in a straight line.

(b) (i) This would assume that the sphere started from rest.
   (ii) Using the equation
   $y = ut + \tfrac{1}{2}at^2$
   dividing through by t
   $\dfrac{y}{t} = u + \tfrac{1}{2}at$
   (iii) (1) When $t = 0$, $u = \dfrac{y}{t}$, the intercept
      $u = 2.8\,\mathrm{m\,s^{-1}}$

(2) Gradient of graph $= \dfrac{4.7 - 2.8}{0.4} = 4.75$

Acceleration $= 2 \times$ gradient $= 9.5\,\mathrm{m\,s^{-2}}$

(3) $v^2 = 2gh$

$2.8^2 = 2 \times 9.8 \times h$

$h = 0.40\,\mathrm{m}$

🄴 7 marks. In part (i) the candidate probably knew the answer, but has not explained it clearly. The sphere does start from rest, but the point to be made is that it assumes the timing starts when the sphere is at rest. Part (ii) is explained clearly. Explanation is very important here because the result is given in the question and it is the derivation that is required. Parts (iii) (1) and (2) are correct and explained well. In part (iii) (3) the method is correct but the candidate has used $9.8\,\mathrm{m\,s^{-2}}$ as the value for g, rather than the value found in part (iii) (2).

■ ■ ■

Answer to Question 2.3: candidate B

(a) (i) u is the initial speed,

s is the distance,

t is the time,

a is the acceleration.

(ii) The acceleration is constant.

🄴 2 marks. In part (i) the candidate has not made it clear that s is the distance moved in a time t. In part (ii), it has not been mentioned that the equation applies to motion in a straight line.

(b) (i) The sphere does not start at L_1.

(ii) Dividing by t

$\dfrac{y}{t} = u + \tfrac{1}{2}at$

(iii) (1) Intercept $= 2.7$

$u = 2.7\,\mathrm{m\,s^{-1}}$

(2) Gradient of graph $= 4.9$

Acceleration $= 2 \times$ gradient $= 9.8\,\mathrm{m\,s^{-2}}$

(3) $v^2 = 2gh$

$2.8^2 = 2 \times 9.8 \times h$

$h = 0.40\,\mathrm{m}$

🄴 4 marks. In part (i) the candidate has given insufficient explanation. The sphere does not start at L_1, but the point here is that the graph would yield a straight line only when the timing starts with the sphere at rest. Part (ii) is not explained clearly and does not score a mark. The starting equation should be given. In part (iii) (1) the intercept has been read wrongly, and in part (iii) (2) marks have been lost because there is no working to show how the gradient was obtained, and the value is incorrect. In part (iii) (3) the calculation is correct, when based on the candidate's answer to part (iii) (2), and is given full marks.

Question 2.4

(a) (i) State an equation relating the force F acting on an object of mass m to its acceleration a.
(ii) List the conditions necessary for the equation in (i) to apply. (4 marks)

(b) A trolley of mass 820 g and weight 8.0 N accelerates down a slope at an angle of 25° to the horizontal, as illustrated below.

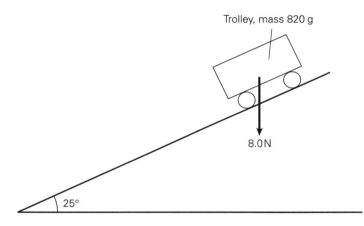

Calculate:
(i) the component of the weight down the slope
(ii) the acceleration of the trolley to be expected from the force in (i) (4 marks)

(c) The acceleration of the trolley is measured and found to be 3.9 m s⁻².
(i) Suggest why this value is different from the value found in (b) (ii).
(ii) Calculate the angle of the slope to the horizontal such that the trolley would move down it at constant speed. (4 marks)

Total: 12 marks

Answer to Question 2.4: candidate A

(a) (i) $F = ma$
(ii) Mass is constant and the force and acceleration are in the same direction.

3 marks. The equation is correct, but in part (ii) the candidate has not mentioned that F is the resultant force acting on the mass.

(b) (i) Weight down slope = $8.0\sin25 = 3.38\,\text{N}$

(ii) $F = ma$

$3.38 = 0.82 \times a$

$a = 4.12\,\text{m s}^{-2}$

 4 marks. The calculations are correct and adequate explanation has been given.

(c) (i) Actual acceleration is less, so there could be some friction acting up the slope.

(ii) For acceleration of $3.9\,\text{m s}^{-2}$, force is $0.82 \times 3.9 = 3.2\,\text{N}$.

Friction force = $3.38 - 3.2 = 0.18\,\text{N}$

This is the component of weight down the slope

$0.18 = 8.0\sin\theta$

$\theta = 1.3°$

 4 marks. The candidate has stated the direction in which the frictional force acts. The calculation is correct.

■ ■ ■

Answer to Question 2.4: candidate B

(a) (i) $F = ma$

(ii) Mass is constant.

 2 marks. The equation is correct, but in part (ii) the candidate has not mentioned that F is the resultant force acting on the mass and that the force and acceleration are in the same direction.

(b) (i) Weight down slope = $8.0\cos65 = 3.38\,\text{N}$

(ii) $F = ma$

$3.38 = 0.82 \times a$

$a = 4.12\,\text{m s}^{-2}$

 4 marks. The calculations are correct and adequate explanation has been given.

(c) (ii) There must be some friction.

(ii) Force accelerating trolley = $0.82 \times 3.9 = 3.2\,\text{N}$

Friction force = $3.38 - 3.2 = 0.18\,\text{N}$

This is the component of weight down the slope

$0.18 = 8.0\cos\theta$

$\theta = 88.7°$

 2 marks. In part (i) the answer 'friction' is insufficient for AS/A-level. Frictional forces have direction. In part (ii) the angle to the vertical has been found. The candidate should have looked at the answer and realised that the angle is far too large. Perhaps the mistake would then have been obvious and another mark would have been scored.

Question 2.5

A ball is thrown horizontally with a speed of $6.0\,\mathrm{m\,s^{-1}}$ from the top of a building of height 11.0 m, as illustrated below.

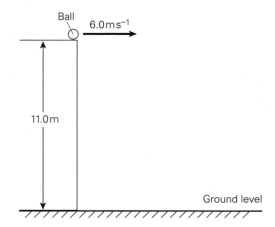

It may be assumed that air resistance is negligible and that the acceleration of free fall is $9.8\,\mathrm{m\,s^{-2}}$.

(a) Use Newton's laws to explain why the horizontal velocity remains constant and why the vertical velocity changes. (4 marks)

(b) Determine:
  (i) the time taken for the ball to reach ground level
  (ii) the horizontal distance travelled by the ball (4 marks)

(c) (i) Draw a sketch to show the path of the ball from the time it is released until it hits the ground. Label this path N.
  (ii) On the same sketch, draw the path of the ball assuming that air resistance is not negligible. Label this path A. (4 marks)

Total: 12 marks

■ ■ ■

Answer to Question 2.5: candidate A

(a) Newton's first law states that a body stays at constant velocity unless acted upon by a force. In the horizontal direction, there is no force and so horizontal velocity is constant. In the vertical direction, weight acts downwards and so by Newton's second law, $F = ma$, and the body will accelerate downwards.

> 🄴 4 marks. This is a good answer. Newton's first and second laws have been stated and their consequences explained. Although full statements of the laws have not been given, they are adequate for this application.

(b) (i) $s = ut + \frac{1}{2}at^2$

$11 = \frac{1}{2} \times 9.8 \times t^2$

$t = 1.5\,\text{s}$

(ii) Distance from building $= 6 \times 1.5 = 9.0\,\text{m}$

> 4 marks. The calculations are correct, with adequate explanation.

(c)

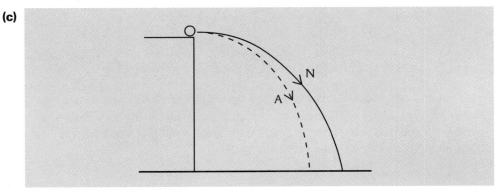

> 4 marks. The candidate was asked to 'draw a sketch'. In this case, the candidate took care to include the main features. In part (i) the ball starts off horizontally and the path has the correct curvature. It is not expected that any points would be plotted, but the path should not be an arc of a circle. In part (ii) the path is shown 'inside' the path N and it hits the ground at an angle closer to the vertical.

■ ■ ■

Answer to Question 2.5: candidate B

(a) Horizontally there is no force and so by Newton's first law, the horizontal velocity is constant. In the vertical direction, the weight makes the sphere accelerate due to the second law.

> 1 mark. The laws have not been stated and the direction of the acceleration in the vertical plane has not been made clear.

(b) (i) $6 = 4.9 \times t^2$

$t = 1.1\,\text{s}$

(ii) Distance $= 6 \times 1.1 = 6.6\,\text{m}$

> 2 marks. The calculation in part (i) is incorrect. It may be that the candidate has substituted a wrong value for s. On the other hand, the candidate may think that the equation is $v = \frac{1}{2}at^2$. Because the candidate has made a substitution error without giving the equation, all the marks have been lost. Always give the equations in symbols before substituting. Part (ii) is correct when based on the answer to part (i).

(c)

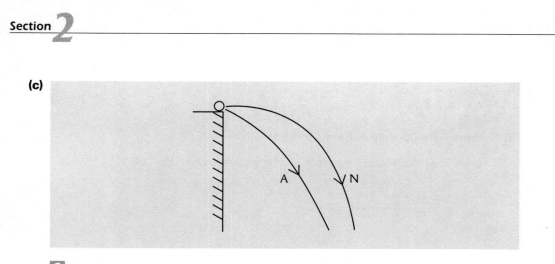

2 marks — generously. In part (i) the ball travels horizontally for some distance after leaving the building. The curved section of path N is satisfactory. Path A is shown 'inside' the path N but its shape is incorrect and it hits the ground at an angle to the vertical that is larger than for path N.

Current electricity

Question 3.1

(a) Explain what is meant by the *potential difference* (pd) between two points.

(2 marks)

(b) A battery of emf 12.0 V and negligible internal resistance is connected in series with two resistors of resistance 2000 Ω and 6000 Ω as shown in the diagram.

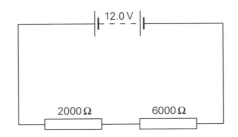

Calculate:
(i) the current in the circuit
(ii) the pd across the 2000 Ω resistor

(4 marks)

(c) The 6000 Ω resistor in the diagram is replaced with a thermistor. State and explain what will happen to the potential difference across the 2000 Ω resistor as the thermistor heats up.

(3 marks)

Total: 9 marks

■ ■ ■

Answer to Question 3.1: candidate A

(a) pd = energy ÷ charge

The energy is converted from electrical to some other form as the charge flows between the points.

   📧 2 marks — a good answer. The ratio of energy to charge is clear and the distinction between pd and emf (i.e. the direction of energy transfer) has been made.

(b) (i) emf = current × total resistance
$$12.0 = I \times (6000 + 2000)$$
$$I = 1.50 \, \text{mA}$$
(ii) pd $= 1.50 \times 10^{-3} \times 2000$
$$= 3.00 \, \text{V}$$

> 4 marks. The equations for emf and for pd in terms of current and resistance are clear, as is the expression for resistors in series. Answers have been given (quite correctly) to 3 significant figures, although two would have been acceptable here.

(c) As temperature rises, the resistance of the thermistor decreases. The current in the circuit will increase and so the pd across the $2000\,\Omega$ resistor ($V = IR$) will rise.

> 3 marks. The effect of temperature change has been stated and the consequence on the pd clearly explained. Note that the candidate has stated 'increased' or 'decreased', rather than 'changed'. Some candidates are reluctant to commit themselves to stating in which direction changes take place. They then lose credit.

■ ■ ■

Answer to Question 3.1: candidate B

(a) The work done moving unit charge from one point to the other.

> No marks. The ratio of $\dfrac{\text{work done}}{\text{charge}}$ is not clear and no detail of the energy transfer has been given.

(b) (i) $12 = I \times 8000$
$I = 1.5 \times 10^{-3}\,\text{mA}$
(ii) pd $= 1.5 \times 10^{-3} \times 2000$
$= 3.0\,\text{V}$

> 3 marks. The explanation is barely adequate and the candidate has relied on correct values to indicate what equations have been used. This is bad practice since any error in a value would mean a serious loss of marks. Always write down relevant equations. One mark has been lost for the answer to (i) because both the prefix 'm' and the power of ten have been given. Be careful not to do that.

(c) The thermistor's resistance decreases and the pd across the $2000\,\Omega$ resistor changes.

> 1 mark — given generously. The candidate has implied that the temperature rises (see the question). However, this is a typical answer where the conditions for a change, and the direction of any subsequent changes, have not been detailed.

Question 3.2

The 'lead' in a pencil is a non-metallic material (a form of carbon) having a resistivity at room temperature of $4.3 \times 10^{-3}\,\Omega\text{m}$. In one such pencil, the 'lead' is of length 12 cm and area of cross-section $1.5 \times 10^{-6}\,\text{m}^2$.

(a) (i) Show that the resistance of the 'lead' is $340\,\Omega$ at room temperature.
(ii) Suggest, with a reason, what change, if any, will occur in the resistivity of the 'lead' when its temperature rises.

(5 marks)

(b) A line of length 25 cm and width 1.5 mm is drawn with the pencil on a sheet of paper. The line is found to have a resistance of 1.4 MΩ.
   (i) Calculate the thickness of the line.
   (ii) Explain what can be deduced about the size of a molecule of the material of the pencil lead.   (5 marks)

Total: 10 marks

Answer to Question 3.2: candidate A

(a) (i) $R = \dfrac{\rho L}{A}$

$$= \dfrac{4.3 \times 10^{-3} \times \dfrac{12}{100}}{1.5 \times 10^{-6}}$$

$$= 344\,\Omega$$

(ii) As the temperature rises, the resistance will decrease because the material is a non-metal. Assuming that the area of cross-section and the length do not change much, then the resistivity will decrease.

🄔 5 marks — a good answer. In part (i) the formula has been stated clearly and the substitution, particularly for L in metres, has been shown. Note that there is no mark for the answer — it has been given in the question! The marks are for clear explanation as to how the answer is obtained. In part (ii) the change in resistance with temperature has been made clear. Other factors that could affect the resistivity have been considered and a valid conclusion has been reached. Note that there is no unique answer to part (ii). Since the temperature coefficient of the material has not been specified, any valid argument that includes all the variables on which resistivity depends would be given full credit.

(b) (i) $A = \dfrac{\rho L}{R}$

$$d \times 1.5 \times 10^{-2} = \dfrac{4.3 \times 10^{-3} \times 0.25}{1.4 \times 10^{6}}$$

$$d = 5.1 \times 10^{-8}\,\text{m}$$

(ii) A molecule must be at least this length because the line must be at least one molecule thick.

🄔 4 marks. In part (i) the equation has been re-arranged correctly; however, there has been an arithmetical error when converting the width from millimetre to metre, so one mark has been lost. Otherwise the calculation is correct. The answer should be $5.1 \times 10^{-7}\,\text{m}$. Part (ii) scores the maximum marks because, although the answer in part (i) is incorrect, the deduction and explanation are perfectly valid.

Answer to Question 3.2: candidate B

(a) (i) $R = \dfrac{\rho L}{L}$

$\qquad = 4.3 \times 10^{-3} \times \dfrac{L}{A}$

$\qquad = 344\,\Omega$

(ii) The resistance changes with temperature and so the resistivity decreases.

> e 1 mark. In part (i) the formula is clear for one mark. However, the working is not given so a possible second mark has been lost. There is no credit in part (ii). Although the resistivity has been said to decrease, the direction of change of resistance has not been given. Furthermore, other factors affecting resistivity have not been included.

(b) (i) $A = \dfrac{\rho R}{L}$

$\qquad d \times 0.15 = 4.3 \times 10^{-3} \times 1.4 \times \dfrac{10^6}{25}$

$\qquad d = 1605\,\text{cm}$

(ii) The molecules are quite large.

> e No marks. In part (i) the manipulation of the formula for resistivity is faulty. Furthermore, the lengths are in centimetre but the resistivity has been left in $\Omega\,\text{m}$. The answer is nonsense. A response like this is not an uncommon sight for examiners. You should always look at your answers and decide whether they are reasonable. Knowing that the thickness of a pencil line is very much less than a millimetre would have alerted you to the fact that there has been a serious error. This error is likely to be in the formula or in powers of ten in the conversion of units. The conclusion in part (ii) would not score any marks. A valid conclusion could be drawn from a wrong answer in (i). However, this conclusion is vague and, in fact, meaningless — 'quite large' in comparison to what?

Question 3.3

(a) (i) State Kirchhoff's first law.

(ii) Explain why Kirchhoff's first law is a consequence of the law of conservation of charge. *(4 marks)*

(b) You are given some resistors, each of resistance $6.0\,\Omega$. Show how you would connect a number of these to produce a resistor having a total resistance of:

(i) $2.0\,\Omega$

(ii) $4.5\,\Omega$ *(3 marks)*

(c) The diagram below shows a complete circuit including two cells of emfs E_1 and E_2. Write down equations, in terms of I_1, I_2, I_R, E_1, E_2, R_1, R_2 and R, for:

(i) the currents at junction A

(ii) the loop ADEFA

(iii) the loop BFECB *(5 marks)*

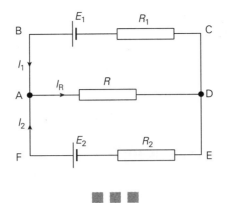

Total: 12 marks

■ ■ ■

Answer to Question 3.3: candidate A

(a) (i) The sum of the currents entering any junction in a circuit is equal to the sum of the currents leaving the junction.

(ii) Current is the rate of flow charge. Since charge is conserved, the charge entering the junction must equal the charge leaving. So, the rates must also be the same.

📧 4 marks. Part (i) is a clear statement of Kirchhoff's first law. In part (ii) the link between Kirchhoff's law and charge conservation is adequately explained.

(b) (i)

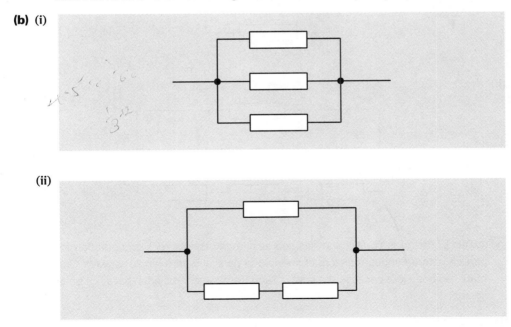

(ii)

📧 2 marks. Part (i) is correct. Possibly 1 mark for part (ii). The configuration is correct, with the input and output leads, but there are only two resistors in series in one arm of the parallel combination (there should be three).

(c) (i) $I_1 + I_2 = I_R$
(ii) $E_2 = (I_1 + I_2)R + I_2R_2$
(iii) $E_1 - E_2 = -I_2R_2 + I_1R_1$

e 5 marks. In part (ii) the candidate has substituted $(I_1 + I_2)$ for I_R. This was unnecessary for full credit to be awarded.

■ ■ ■

Answer to Question 3.3: candidate B

(a) (i) The current going into a junction equals the current leaving the junction.
(ii) Current is the rate of flow of charge.

e 2 marks only. In part (i) there is no mention of the sum of currents. In part (ii) the concept of current as flow of charge has been given. However, the link between Kirchhoff's law and charge conservation has not been made.

(b) (i)

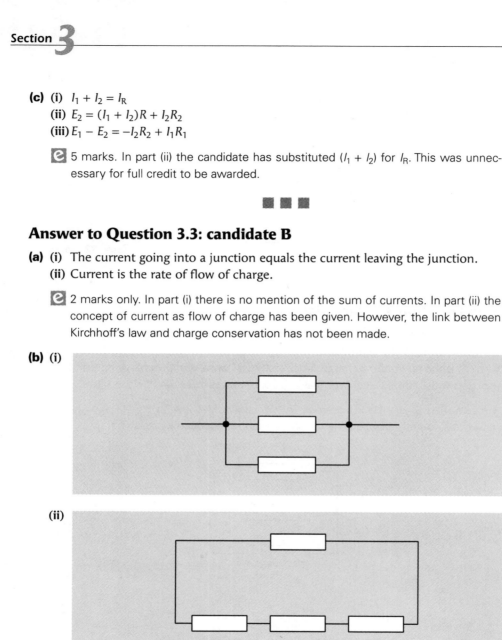

(ii)

e 2 marks. Part (i) is correct and has scored 1 mark. The correct configuration for part (ii) is three series resistors connected in parallel with a single resistor. One mark has been lost because the leads to and from the combination have not been shown.

(c) (i) $I_1 + I_2 = I_R$
(ii) $E_2 = I_R R + I_2 R_2$
(iii) $E_1 + E_2 = I_2 R_2 + I_1 R_1$

Question 3.4

(a) The graph below shows the variation with potential difference of the current in a component X.

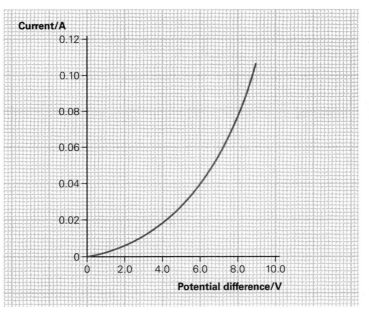

(i) Describe qualitatively how the resistance of X varies with applied potential difference.

(ii) Determine the resistance of X for a potential difference of 5.0 V.   (4 marks)

(b) Component X is connected into the circuit shown below.

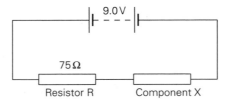

The battery has an emf of 9.0 V and negligible internal resistance. R is a fixed resistor of resistance 75 Ω.

(i) Calculate the current in the resistor R when the potential difference across it is 4.5 V.

(ii) Copy the graph from part **(a)** and on it show the variation with potential difference across R of the current in R.

(iii) Use the graph lines in (ii) to determine the current in the circuit illustrated.

(6 marks)

(c) The fixed resistor R in the circuit in **(b)** is to be changed so that the current in the circuit is 0.060 A. Determine the new value of the resistance of R. (3 marks)

Total: 13 marks

Answer to Question 3.4: candidate A

(a) (i) The resistance decreases as the potential difference increases.

(ii) When $V = 5.0$ V, current $= 0.027$ A.

$$R = \frac{V}{I} = 185\,\Omega$$

🔴 4 marks. The candidate has given the direction of each change and the calculation is correct.

(b) (i) $I = \dfrac{V}{R} = \dfrac{4.5}{75} = 0.060$ A

(ii)

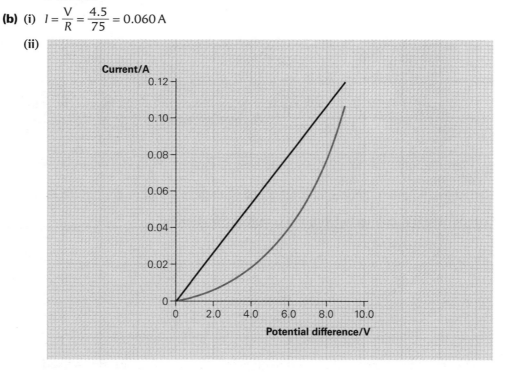

(iii) Same current in X as in R and potential differences must add up to 9.0 V. This occurs when current is 0.040 A.

e 6 marks. The graph is correct and the work has been explained well.

(c) When the current is 0.06 A, the pd across X is 7.2 V.
So, pd across R = 9.0 − 7.2 = 1.8 V

Resistance of R = $\dfrac{1.8}{0.06}$ = 30 Ω

e 3 marks. The trick is to remember that the straight line in the graph of part (b) (ii) no longer applies but that the pd across X and across R must add up to 9.0 V.

■ ■ ■

Answer to Question 3.4: candidate B

(a) **(i)** The resistance changes as the voltage increases.
(ii) Resistance is found from the gradient of the graph.

Gradient = $\dfrac{0.097}{6.7}$ = 0.0145

Resistance = $\dfrac{1}{0.0145}$ = 69 Ω

e 1 mark. In part (i) the candidate realises that the resistance changes, but has not said in which direction. Resistance is given as V/I, but the gradient has been found. It is important to realise that resistance is the ratio of pd and current at a point on the graph, not the gradient.

(b) **(i)** $I = \dfrac{4.5}{75} = 0.060\,\text{A}$

(ii)

(iii) $I = \dfrac{9}{R+X} = \dfrac{V}{75}$

🅔 3 marks. Parts (i) and (ii) are correct. However, the candidate does not appear to understand what has to be done. Always read the question carefully. The clue was there — *'use the graph lines in (ii) ....'*

(c) For a current of 0.06 A, pd across X = 7.1 V.
Resistance of X = 118 Ω

Resistance of circuit $= \dfrac{9}{0.06} = 150\,\Omega$

Resistance of R = 150 − 118 = 32 Ω

🅔 2 marks. The method for the calculation is correct. However, the candidate has misread the graph: this is a common error when using 2 mm square graph paper.

Question 3.5

(a) A battery of emf 3.0 V and negligible internal resistance is connected in series with a resistor of resistance 5.0 Ω and a wire AB of length 150 cm, as shown below.

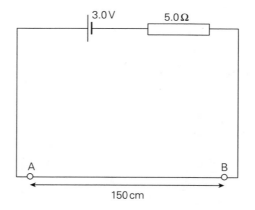

The wire has a cross-sectional area of $3.6 \times 10^{-8}\,m^2$ and the material of the wire has resistivity $4.8 \times 10^{-7}\,\Omega\,m$.
(i) Calculate the resistance of the wire AB.
(ii) Determine:
  (1) the potential difference (pd) across wire AB
  (2) the pd per unit length of AB   (5 marks)

(b) A cell C of emf 1.5 V and a resistor of resistance 10 Ω are connected to the circuit in part **(a)** as shown below.

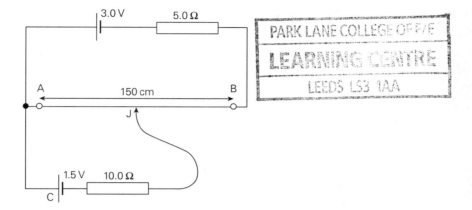

Connection may be made to the bare wire AB by means of a contact J.

(i) Using your answer to **(a)** (ii) (2), determine the position of J on wire AB so that there is no current in the 10 Ω resistor.

(ii) The contact is now moved closer to end B of the wire. State the direction of the current in the 10 Ω resistor. (3 marks)

(c) Suggest two ways in which the circuit illustrated in **(b)** could be changed so that the position of the contact J on wire AB for no current in the 10 Ω resistor is nearer to end B. (2 marks)

Total: 10 marks

■ ■ ■

Answer to Question 3.5: candidate A

(a) (i) $R = \dfrac{\rho L}{A}$

$$= \dfrac{4.8 \times 10^{-7} \times 1.50}{3.6 \times 10^{-8}}$$

$$= 20 \, \Omega$$

(ii) **(1)** pd $= \dfrac{20}{25} \times 3.0 = 2.4 \, \text{V}$

(2) pd per unit length $= \dfrac{2.4}{150} = 0.016 \, \text{V cm}^{-1}$

🄴 5 marks. The candidate has used the potential divider formula quite correctly in part (ii) (1). Be careful — there are many possible mistakes to be made here, particularly when the work is not fully explained.

(b) (i) Needs 1.5 V along the wire.

Balance point $= \dfrac{1.5}{0.016} = 94 \, \text{cm}$

(ii) Current from C to J.

🄴 2 marks. A balance point cannot be just given as '94 cm'. The candidate has failed to give the position. The correct answer would be '94 cm from A'. The answer in part (ii) is correct. No explanation was required. The explanation is that the pd along AB would be greater than the emf of the cell C, thus driving current through cell C.

(c) Increase the value of the 10 Ω and the 5 Ω resistors.

> 1 mark. Changing the value of the 10 Ω resistor would have no effect on the position since, when J is positioned correctly, there is no current in this resistor. The other alternative is to reduce the emf of the 3.0 V battery.

Answer to Question 3.5: candidate B

(a) (i) $R = \dfrac{\rho L}{A}$

$$= \frac{4.8 \times 10^{-7} \times 150}{3.6 \times 10^{-8}}$$

$$= 2000 \, \Omega$$

(ii) (1) Current in circuit $= \dfrac{3}{2005}$

pd across wire $= 2000 \times \dfrac{3}{2005} = 2.99 \, V$

(2) pd per cm of wire $= \dfrac{2.99}{150} = 0.020 \, V \, cm^{-1}$

> 4 marks. In part (i) the method is correct, but the length was left in centimetres, rather than converted to metres. The candidate did not use the potential divider formula in part (ii) but the work is correct, based on the answer to part (i). When in any doubt about the potential divider formula, it is better to work from first principles.

(b) (i) Needs 1.5 V down the wire
Length $= 1.5 \times 0.02 = 0.03 \, cm$

(ii) From J to C

> No marks. The length has not been calculated correctly. The candidate should have realised this when checking whether the answer is sensible. There is a common mistake in part (ii) in that some candidates think the conventional current will always be out of the positive terminal of the battery.

(c) Change the value of the 5 Ω resistor and the emf of the battery.

> No marks. This is a shame; the candidate realises which components must be changed but has failed to give the direction of these changes.

Question 3.6

(a) Distinguish between the electromotive force (emf) of a cell and the potential difference (pd) across a resistor. (3 marks)

(b) A battery of emf 6.0 V and internal resistance 2.0 Ω is connected in series with a variable resistor and an ammeter, as shown below.

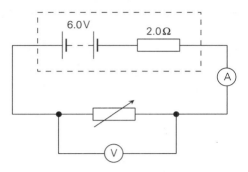

A voltmeter is connected in parallel with the variable resistor. Readings of the current in the circuit are found for different values of the potential difference. These are shown in the following table.

Potential difference/V	Current/A	Resistance/Ω	Power/W
5.00	0.50		
4.00	1.00		
3.00	1.50		
2.00	2.00		
1.00	2.50		

(i) Copy and complete the table.
(ii) Using the values obtained in (i), plot a graph to show the variation with external resistance of the power dissipated in the external resistance.
(iii) Hence determine the value of external resistance at which the power dissipation is a maximum. (7 marks)

(c) Suggest two further values of external resistance at which readings could be taken in order to improve the reliability of your result in **(b)** (iii). (2 marks)

(d) When starting a car, the current from the car battery is approximately 100 A. Suggest why the internal resistance of the battery should be as low as possible. (2 marks)

Total: 14 marks

■ ■ ■

Answer to Question 3.6: candidate A

(a) Both are energy/charge. With emf, the energy is converted into electrical energy whereas for pd, the electrical energy is converted into some other form.

3 marks. It would have been better if the candidate had stated that, for emf, the energy is transferred from some other form into electrical energy.

(b) (i)

Potential difference/V	Current/A	Resistance/Ω	Power/W
5.00	0.50	10.0	2.5
4.00	1.00	4.0	4.0
3.00	1.50	2.0	4.5
2.00	2.00	1.0	4.0
1.00	2.50	0.4	2.5

(ii)

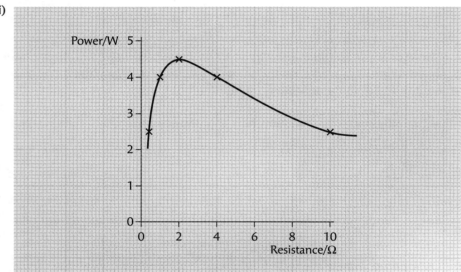

(iii) Value of resistance = 2.0 Ω

 7 marks. The values in part (i) are correct and have been well plotted in part (ii). It is always worth being careful with graph plotting because these marks are easy to score. Remember that graph plotting is not sketching. Follow the same rules as for graph plotting in the practical assessment.

(c) The points should be at resistances of 1.5 Ω and 2.5 Ω.

 2 marks. Yes, the points should be near to the maximum.

(d) So that the battery is not heated too much.

 1 mark. This is true. The other reason is that a large internal resistance would mean that the terminal potential difference would be much less than the emf, thus reducing the current to the starter.

Answer to Question 3.6: candidate B

(a) The emf of a battery puts energy into a circuit. A pd shows that electrical energy is being changed into heat.

> 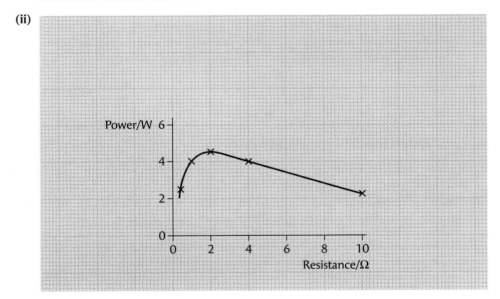 1 mark. A general statement that both are measured as the ratio of energy and charge has not been given. The energy transfers are either incomplete or too specific. Heat is not the only form of energy into which electrical energy is transferred.

(b) (i)

Potential difference/V	Current/A	Resistance/Ω	Power/W
5.00	0.50	10.0	2.5
4.00	1.00	4.0	4.0
3.00	1.50	2.0	4.5
2.00	2.00	1.0	4.0
1.00	2.50	0.4	2.5

(ii)

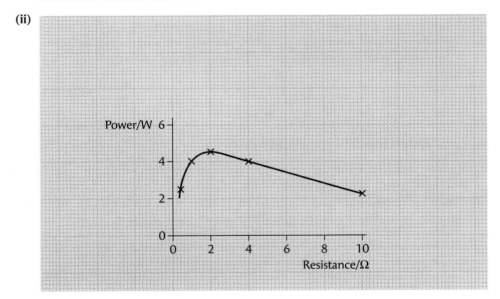

(iii) Resistance = 2 Ω

> 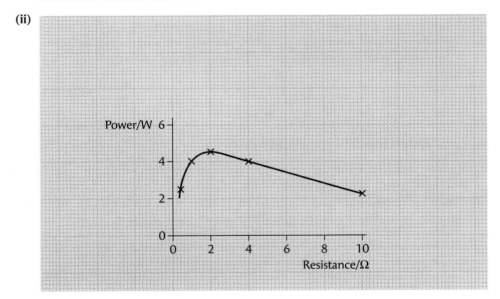 5 marks. The values in part (i) are correct. In part (ii) *the scale for the graph on the y-axis could be doubled.* Also, there is no justification for drawing a straight line for values of resistance greater than 3 Ω.

(c) The points should be at resistances of 6 Ω and 8 Ω.

> No marks. This graph is a typical example of where it is not beneficial to have points equally spaced. It is the region of the maximum that is of importance here.

(d) So that voltage is not lost in the battery.

> 🄴 1 mark. This is poorly expressed but it does convey the idea that the terminal potential difference would be less than the emf. Heating within the battery has not been included.

Question 3.7

(a) Explain:
   (i) what is meant by a good conductor of electricity
   (ii) why metals are classed as good conductors   (4 marks)

(b) The molar mass of copper is 63.5 g and its density is 8900 kg m^{-3}. Show that the number of copper atoms per unit volume is 8.44×10^{28} m^{-3}. The Avogadro constant = 6.02×10^{23} mol^{-1}.   (2 marks)

(c) A copper wire has a diameter of 1.0 mm and carries a current of 2.5 A. It may be assumed that copper has one free electron per atom and that the charge on an electron is 1.6×10^{-19} C.
   (i) Determine the drift speed of the electrons in the wire.
   (ii) Comment on your answer to (i) with reference to the observation that a lamp lights up at almost the same instant as the switch in the circuit is closed.   (4 marks)

Total: 10 marks

■ ■ ■

Answer to Question 3.7: candidate A

(a) (i) A good conductor of electricity is one which has a low resistance and the current in it is large even when the pd across it is small.
   (ii) Metals are good conductors because they contain many free electrons. When the electrons move, they carry charge and moving charge is current.

> 🄴 4 marks. An excellent answer. The candidate has stated what is meant by a good conductor and included what constitutes an electric current.

(b) Mass of 1.0 m^3 = 8900 kg

number of moles in 1 m^3 = $\dfrac{8900}{0.0635}$ = 1.40×10^5

number of atoms in 1 m^3 = $1.40 \times 10^5 \times 6.02 \times 10^{23}$ = 8.44×10^{28}

> 🄴 2 marks. The answer is correct. However, the answer was given in the question, so the marks are awarded for a clearly explained correct method.

(c) (i) $I = nAqv$

$2.5 = 8.44 \times 10^{28} \times \pi \times (1.0 \times 10^{-3})^2 \times 1.6 \times 10^{-19} \times v$

$v = 5.9 \times 10^{-5}\,\text{m s}^{-1}$

(ii) When the current is switched on, all the electrons begin to shuffle along at the same time. The shuffling speed is the drift speed.

@ 2 marks. In part (i) the candidate has made a common error. The diameter was used instead of the radius in the formula for cross-sectional area. The correct answer is $2.36 \times 10^{-4}\,\mathrm{m\,s^{-1}}$. In part (ii) the signal to make the electrons drift along the wire travels at the speed of electromagnetic waves in the wire. This point has not been made clearly.

Answer to Question 3.7: candidate B

(a) (i) A good conductor is one that allows a large current to pass through it.
(ii) Metals are good conductors because they contain electrons that carry the current.

@ No marks. In part (i) the candidate has not stated what is meant by a good conductor. If the potential difference is large enough, then any material can be made to conduct a large current. The important point that there are *free* electrons in a metal has been omitted. Both conductors and insulators contain electrons!

(b) 8900 kg of copper is held in $1.0\,\mathrm{m^3}$.

Number of atoms in $1\,\mathrm{m^3} = \dfrac{8900}{63.5} \times 6.02 \times 10^{23} = 8.44 \times 10^{28}$

@ 1 mark. The method of carrying out the calculation is correct. However, the answer given was not obtained from the candidate's substitution. The molar mass was not converted to kilograms.

(c) (i) $I = nAqv$
$2.5 = 8.44 \times 10^{28} \times \pi \times 0.5^2 \times 1.6 \times 10^{-19} \times v$
$v = 2.36 \times 10^{-10}\,\mathrm{m\,s^{-1}}$
(ii) When the current is switched on, the electrons push each other around at the drift speed.

@ 1 mark. In part (i) the candidate has not converted the cross-sectional area to (metres)^2. Although drift speeds are low, the candidate should have realised this is far too small — about one atomic diameter per second! The error was likely to be in converting units. In part (ii) it has not been explained why all the electrons begin to move at about the same time.

Energy

Question 4.1

(a) (i) Define *power*.

(ii) Starting with an expression for work done, derive a relation for the constant power P required to move against a force F at speed v. (4 marks)

(b) A student carries out an experiment with a toy electric motor in order to determine its useful power output and its efficiency. A mass of 160 g is attached to a thread, the other end of which is tied to the axle of the motor, as shown in the diagram.

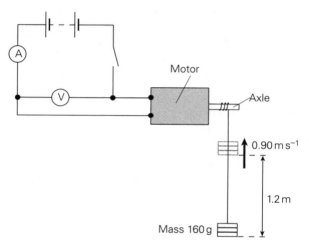

The motor is switched on and the mass is raised, from rest, through a vertical height of 1.2 m in 1.5 s. The mass attains a speed of $0.90\,\text{m s}^{-1}$. During this time, the readings on the voltmeter and the ammeter are 2.5 V and 1.1 A respectively. The acceleration of free fall is $9.8\,\text{m s}^{-2}$.

(i) Determine the average output power of the motor.

(ii) Calculate the input power to the motor.

(iii) Hence determine the efficiency of the motor. (9 marks)

Total: 13 marks

Answer to Question 4.1: candidate A

(a) (i) Power is the rate of doing work.

(ii) Work done = force × displacement in direction of force
$$W = F \times s$$

Dividing both sides by time

$$\frac{W}{t} = F \times \frac{s}{t}$$

But $\frac{W}{t}$ is power and $\frac{s}{t}$ is speed

$$P = Fv$$

e 4 marks. The definition in part (i) is satisfactory. In part (ii) it would have been better to state that 'time' is the time to move the distance s. However, full credit has been given.

(b) (i) Work done $= mgh + \frac{1}{2}mv^2$

$\qquad\qquad\quad = (0.16 \times 9.8 \times 1.2) + (\frac{1}{2} \times 0.16 \times 0.9^2)$

$\qquad\qquad\quad = 1.95\,J$

Power $= \dfrac{1.95}{1.5} = 1.3\,W$

(ii) Electric power $= VI = 2.5 \times 1.1 = 2.75\,W$

(iii) Efficiency $= \dfrac{1.3}{2.75} = 47\%$

e 9 marks. The candidate has explained the work well. It is always advisable to give equations before substitution so that, if a mistake is made in the working, some marks will be scored.

■ ■ ■

Answer to Question 4.1: candidate B

(a) (i) Power is the work done in unit time.

(ii) Work done $= F \times s$

Dividing by time

$$\frac{work}{t} = F \times \frac{s}{t}$$

But $\dfrac{work}{time}$ is power

$$P = Fv$$

e 2 marks. The definition in part (i) is not satisfactory in that it has not been made clear that power is the *ratio* of work done and time taken. In part (ii) no explanation is given for the symbol s, $\frac{s}{t}$ or what is meant by 'time'.

(b) (i) Work done $= mgh$

$\qquad\qquad\quad = 0.16 \times 9.8 \times 1.2$

$\qquad\qquad\quad = 1.88\,J$

Power $= \dfrac{1.88}{1.5} = 1.25\,W$

(ii) Electric power $= 2.5 \times 1.1 = 2.75\,W$

(iii) Efficiency $= \dfrac{1.25}{2.75} = 45\%$

7 marks. Explanation is adequate. However, the candidate has failed to include the gain in kinetic energy when calculating the work done by the motor. The remainder of the work is correct when the candidate's value for work done is assumed.

Question 4.2

(a) Define *work done* by a force.   (2 marks)

(b) An object of mass m is moving in a straight line. Its speed is increased from u to v by a resultant force F while the object moves a distance s.
  (i)   Derive an expression for the work done by the force F in terms of m, v and u.
  (ii) Hence derive an expression for the kinetic energy of an object travelling at speed v.   (6 marks)

(c) Suggest and explain two reasons why a car uses more fuel when travelling a certain distance at high speed than at low speed.   (4 marks)

Total: 12 marks

Answer to Question 4.2: candidate A

(a) It is the force multiplied by the distance moved by the force in the direction of the force.

   2 marks. Well defined. The product is clear, as are the relative directions of the force and the displacement.

(b) (i) $v^2 = u^2 + 2as$ and $F = ma$
  Combining these equations, $v^2 = u^2 + \dfrac{2Fs}{m}$
  Therefore work done $= Fs = \frac{1}{2}mv^2 - \frac{1}{2}mu^2$
  (ii) Fs is energy and so the other terms must also be energy.
  Kinetic energy $= \frac{1}{2}mv^2$

   5 marks. Many candidates lose several marks in a derivation such as this because they do not give sufficient explanation. Part (i) has been answered very clearly. In part (ii), the identification of the term $\frac{1}{2}mv^2$ as being energy is clear. However, the candidate should have said that the only quantity that varies in this term is speed and therefore $\frac{1}{2}mv^2$ must be energy associated with speed.

(c) At higher speeds the car has more kinetic energy. This is provided by the fuel and is not got back when the car stops. Also, there is more friction and so more work is done, so more fuel is used.

   3 marks. Remember that this is AS/A-level. A blank reference to 'friction' is insufficient. The candidate should have given the source of this friction, i.e. air resistance (or drag).

Answer to Question 4.2: candidate B

(a) It is the force times distance.

> e No marks. The candidate has not distinguished between 'distance from' and 'distance moved'. Also, there is no reference to the relative directions of force and 'distance'.

(b) (i) $v^2 = u^2 + 2as$
$F = ma$
$Fs = \frac{1}{2}mv^2 - \frac{1}{2}mu^2$
(ii) Fs is work done and so
kinetic energy $= \frac{1}{2}mv^2$

> e 3 marks. Part (i) is not fully answered until part (ii) is reached. In part (i) the candidate did not identify Fs with work done. In part (ii) the identification of the term $\frac{1}{2}mv^2$ as being energy has not been made. Furthermore, there is no explanation as to why the term $\frac{1}{2}mv^2$ should be energy associated with speed.

(c) At higher speeds the car has more kinetic energy and there is more friction so more fuel is used.

> e 1 mark. The candidate has identified one factor but the reference to friction is inadequate at AS/A-level. The explanations have not been provided.

Question 4.3

(a) Explain what is meant by *potential energy*. (2 marks)

(b) The passenger carriage on a fairground ride has a mass of 500 kg. The carriage starts at a height of 30 m above the ground and, after passing through a dip, it finishes at a height of 25 m above the ground, as illustrated below.

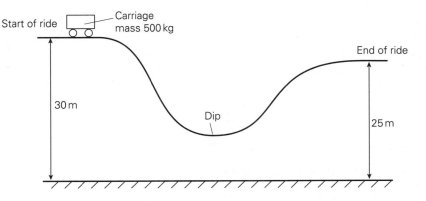

(i) Calculate the change in potential energy between the start and the end of the ride. The acceleration of free fall is 9.8 m s^{-2}.

(ii) The carriage starts from rest. Assuming there are no frictional forces acting on the carriage, calculate the speed of the carriage at the end of the ride.

(iii) In practice, the speed of the carriage at the end of the ride is 1.9 m s⁻¹. Determine the energy lost due to frictional forces. (7 marks)

(c) Suggest, with reasons, two effects on the ride if the depth of the dip were to be increased. (4 marks)

Total: 13 marks

Answer to Question 4.3: candidate A

(a) Potential energy is the ability to do work due to the position of the object.

> ⓔ 2 marks. Examples of potential energy were not expected.

(b) (i) Change in potential energy $= mg\Delta h$
$$= 500 \times 9.8 \times 5$$
$$= 24\,500\,\text{J}$$

(ii) $\frac{1}{2}mv^2 = 24\,500$
$v^2 = 98$
$v = 9.9\,\text{m s}^{-1}$

(iii) Kinetic energy at end $= \frac{1}{2} \times 500 \times 1.9^2$
$$= 903\,\text{J}$$

Energy lost $= 23\,600\,\text{J}$

> ⓔ 7 marks. The calculations are correct and the work is explained well.

(c) Final speed would be the same because the change in potential energy is the same. The speed at the bottom of the dip would be greater because the change in p.e. would be greater.

> ⓔ 2 marks. The final speed will, in fact, be less. The candidate is right in saying that the total change in potential energy is the same, but deepening the dip will mean that the frictional forces will act over a longer distance. Thus, more energy is lost due to friction and the final speed will be less. As stated, the speed at the bottom of the dip will be greater.

Answer to Question 4.3: candidate B

(a) Potential energy is energy due to position.

> ⓔ 1 mark. Note that, in the question, both of the words 'potential' and 'energy' are in *italics*. This means that both words must be explained. The candidate has not said what is meant by *energy*.

(b) (i) Change in potential energy $= (500 \times 10 \times 30) - (500 \times 10 \times 25)$
$$= 25\,000\,J$$

(ii) $\frac{1}{2}mv^2 = 25\,000$
$v = 100\,m\,s^{-1}$

(iii) Kinetic energy at end $= \frac{1}{2} \times 500 \times 1.9^2$
$$= 903\,J$$

Energy lost $= 24\,100\,J$

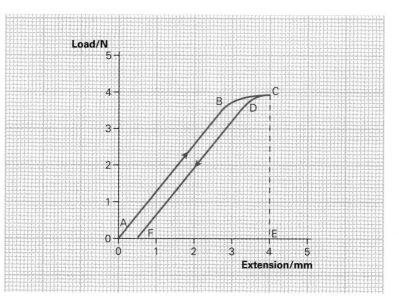 5 marks. In part (i) the candidate did not use the given value for the acceleration of free fall. In part (ii) a common error has been made. The amount of explanation is minimal and the candidate forgot to take the square root. In fact, the candidate should have realised this because the answer is unrealistic.

(c) The speed at the bottom of the dip would be greater but the final speed would be the same.

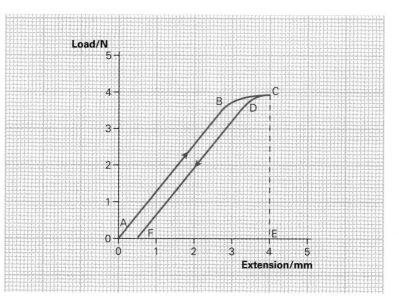 1 mark. Despite the wording of the question, the candidate has given no explanation. The final speed would, in fact, be less.

Question 4.4

(a) (i) Explain what is meant by *strain energy*.
(ii) Show that the strain energy E of a spring having an elastic (spring) constant k and extended by an amount x is given by

$$E = \frac{1}{2}kx^2$$

(5 marks)

(b) The graph below shows the variation with load of the extension of a wire.

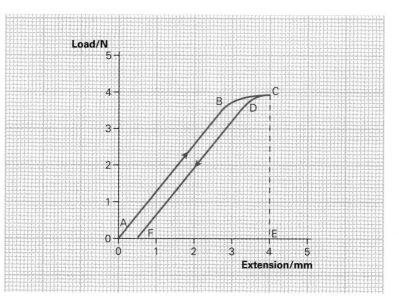

Initially, the extension of the wire is zero. Describe what is represented by:
(i) the area ABCEA
(ii) the area FDCEF
(iii) the area ABCDFA   (4 marks)

(c) (i) Use the graph to determine the energy expended when the load is increased to cause the extension to change from zero to 4.0 mm and then the load is reduced to zero.
(ii) Suggest what happens to the energy in (i).   (5 marks)

Total: 14 marks

■ ■ ■

Answer to Question 4.4: candidate A

(a) (i) Work that can be done due to energy stored in a material when it is deformed.
(ii) Work done = force × extension
$$= \tfrac{1}{2} Fx$$
but $F = kx$
so $E = \tfrac{1}{2} kx^2$

🅮 4 marks. The explanation in part (i) is satisfactory. In part (ii) the candidate has not explained how the factor of $\tfrac{1}{2}$ came into the equation. In fact, the work done is the product of the *average* force and the extension. When the force increases uniformly from zero to a final value F, the average force is $\tfrac{1}{2} F$.

(b) (i) Work done in extending the wire.
(ii) Energy released when wire contracts.
(iii) Energy used to deform the wire.

🅮 3 marks. The candidate should have stated that the energy was used to cause the permanent deformation of the wire. Deformation may be elastic or plastic.

(c) (i) Area enclosed is 190 small squares.
Energy = $190 \times 0.1 \times 0.1 \times 10^{-3}$
$= 0.0019 \, \text{J}$
(ii) Some of the energy is used to change the positions of atoms in crystals. Some causes heating of the wire.

🅮 5 marks. In part (i) it would have been better if the candidate had explained that each small square on the graph represents $(0.1 \times 0.1 \times 10^{-3}) = 1.0 \times 10^{-5} \, \text{J}$, but a mark was not deducted.

■ ■ ■

Answer to Question 4.4: candidate B

(a) (i) Energy stored in a material when it is deformed.

(ii) Work done = force × extension

$$= Fx$$

but $F = kx$

so $E = kx^2$

average energy $= \frac{1}{2}kx^2$

e 2 marks. The explanation in part (i) does not include any indication of what is meant by energy in this context. The candidate did not read the question carefully. The word 'energy' is in italics and should, therefore, have been explained. In part (ii) the candidate does not understand how to introduce the factor of $\frac{1}{2}$. Standard derivations should be learned carefully because they can provide easy marks!

(b) (i) Work done to extend the wire.

(ii) Energy got back when the wire is released.

(iii) Energy lost in stretching.

e 2 marks. Part (iii) is too vague to be given credit. The candidate should have stated that the energy was used to cause the permanent stretch in the wire.

(c) (i) Area enclosed is 170 small squares.

Energy = 17 J

(ii) The wire is heated.

e 1 mark. In part (i) the candidate has not included the tapering part of the area when counting squares. Furthermore, the step from area to energy is incorrect. If the candidate had given some explanation of this stage in the calculation, some marks may have been awarded. It is always dangerous not to explain the working. In part (ii) no mention has been made of the energy required to change permanently the positions of the atoms within the solid.

Thermal physics

Question 5.1

(a) The specific heat capacity of copper is $400\,\text{J}\,\text{kg}^{-1}\,\text{K}^{-1}$.
  (i) Explain the meaning of this statement.
  (ii) On the basis of the kinetic theory, explain what happens to the thermal energy when copper is heated. (4 marks)

(b) On one particular model of car, the braking system consists of a metal disc attached to each of the four wheels. When the brakes are applied, pads made of a poor thermal conductor grip the discs, bringing them to a halt. Each disc is made of steel of specific heat capacity $460\,\text{J}\,\text{kg}^{-1}\,\text{K}^{-1}$ and has a mass of 2.5 kg. The car has mass 950 kg and is travelling at a speed of 80 km h^{-1}. Calculate the rise in temperature of the brake discs when the car is brought to rest, assuming that 75% of the energy of the car is converted into thermal energy in the discs. (5 marks)

(c) Suggest how the discs of the brakes may be cooled. (2 marks)

Total: 11 marks

Answer to Question 5.1: candidate A

(a) (i) It means that 400 J of heat are required to raise the temperature of 1 kg of copper by 1 kelvin.
  (ii) The energy increases the vibrations of the atoms of copper.

    🄔 2 marks. The explanation in part (i) is good. However, in part (ii) the candidate should have said that the *amplitude of vibration* is increased. No mention has been made of the increase in potential energy of the atoms as a result of their increased separation.

(b) $80\,\text{km}\,\text{h}^{-1} = \dfrac{80\,000}{600} = 22.2\,\text{m}\,\text{s}^{-1}$

Kinetic energy $= \frac{1}{2}mv^2 = \frac{1}{2} \times 950 \times 22.2^2$
$$= 2.35 \times 10^5\,\text{J}$$
$0.75 \times 2.35 \times 10^5 = mc\theta = 2.5 \times 460 \times \theta$
$\theta = 153\,\text{K}$

    🄔 4 marks. The explanation is satisfactory and it is clear what the candidate has done. The only error is that only one disc was considered when, in fact, four discs are heated. The answer should be 38 K.

(c) The discs should be cooled by air currents.

1 mark. The candidate should have stated that the discs are in freely moving air so that they are cooled by (forced) convection.

■ ■ ■

Answer to Question 5.1: candidate B

(a) (i) The amount of heat required to heat 1 kg of copper by 1 degree is 400 joules.
(ii) The heat is used to cause vibrations of the atoms.

1 mark only. In part (i) the unit of temperature rise has not been included. In part (ii) any reference to increased potential energy of the atoms has been omitted and the candidate gives the impression that the atoms were not vibrating before heating.

(b) Kinetic energy $= \frac{1}{2} \times 950 \times 80^2 = 3.04 \times 10^6$ J
$0.75 \times 3.04 \times 10^6 = 10 \times 460 \times \theta$
$\theta = 496$ K

4 marks. The explanation given is minimal. It is always wise to quote the formulae used so that, should there be a mistake in the substitution, the examiner will know where the fault lies. The candidate did not convert $km\,h^{-1}$ to $m\,s^{-1}$. Inspection of the answer should have made the candidate realise that there is an error in the working.

(c) The discs are cooled by conduction and convection.

No marks. This is too vague to be accepted at AS/A-level. Candidates would be expected to give some detail of the means by which cooling by convection occurs.

Question 5.2

(a) Define specific latent heat of vaporisation of a liquid. (2 marks)

(b) An electric kettle is rated at 2.4 kW and is filled with a mass of 1.5 kg of water. The specific heat capacity and the specific latent heat of water are $4.20\,kJ\,kg^{-1}\,K^{-1}$ and $2280\,kJ\,kg^{-1}$ respectively. Assuming that all the electrical energy provided is used to heat the water, calculate:
(i) the time taken to raise the temperature of the water from 18 °C to 100 °C
(ii) the mass of water evaporated per second when the water is boiling

(6 marks)

(c) Electrical energy costs 8.5 pence per kilowatt-hour. Calculate the cost of heating the water to its boiling point. (2 marks)

Total: 10 marks

■ ■ ■

Answer to Question 5.2: candidate A

(a) Specific latent heat is the quantity of heat required to convert unit mass of liquid to vapour without any change of temperature.

> 2 marks. It would have been better to state that the 'specific latent heat is *numerically equal to* the quantity of heat …'. However, the definition given here would normally be accepted.

(b) (i) Heat required $= mc\theta = 1.5 \times 4200 \times (100 - 18)$
$$= 516\,600\,J$$
Power $\times$ time $=$ energy
$$time = \frac{516\,600}{2400} = 215\,s$$

(ii) Power $\times$ time $= mL$
$2400 \times 1 = m \times 2280$
$m = 1.05\,kg\,s^{-1}$

> 5 marks. The formulae used at each stage have been quoted. This is always to be advised. In part (ii) the specific latent heat was left in $kJ\,kg^{-1}$ rather than converted to $J\,kg^{-1}$. If the candidate had thought about the answer, then the mistake would have been realised. A kettle-full of water does not boil dry in less than 2 seconds!

(c) Number of kilowatt hours $= 2.4 \times \dfrac{215}{3600} = 0.143$

Cost $= 8.5 \times 0.143 = 1.2$ pence

> 2 marks. The calculation can be followed easily.

Answer to Question 5.2: candidate B

(a) Specific latent heat is equal to the ratio of heat supplied and mass.

> No marks. The word equation for calculating latent heat has been given but this is insufficient. The candidate has not mentioned that the heat is used to convert liquid to vapour or that the change occurs at constant temperature.

(b) (i) Heat required $= 1.5 \times 4200 \times 78$
$$= 491\,400\,J$$
Time $= \dfrac{491\,400}{2400} = 205\,s$

(ii) Power $\times$ time $= mL$
$2.4 \times 1 = m \times 2280$
$m = 1.05 \times 10^{-3}\,kg\,s^{-1}$

> 3 marks. The formula used in part (i) has not been quoted. Consequently, since an incorrect temperature change has been used, no marks can be given for this part.

Part (ii) is correct.

(c) Cost $= 2.4 \times \dfrac{215}{60} \times 8.5 = 73$ pence

🄴 1 mark. A very common error has been made that the candidate should have spotted by considering the answer: the conversion of seconds to hours is faulty.

Matter

Question 6.1

(a) Define:
  (i) tensile stress
  (ii) tensile strain (3 marks)

(b) Describe an experiment to determine the Young modulus of a metal in the form of a long wire. (8 marks)

(c) A thick beam of wood is supported at both ends and a large weight is placed at its centre, causing the beam to bend, as shown in the diagram below.

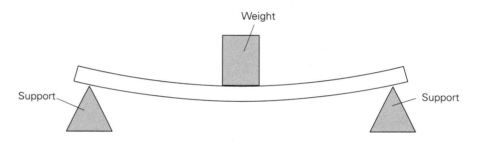

Copy the diagram and on your copy mark a position where the beam is:
  (i) under tension (mark this position T)
  (ii) under compression (mark this position C) (2 marks)
  Total: 13 marks

Answer to Question 6.1: candidate A

(a) **(i)** Stress is force divided by area.
  (ii) Strain is extension divided by original length.

  e 2 marks. The candidate has not stated that the area is the cross-sectional area.

(b) The length l of the test wire is measured and also its diameter, using a micrometer screw gauge. A weight is put on the test wire and the vernier is read. Another weight is added and the new reading of the vernier is taken (see the diagram below). This gives the extension e for a weight W. The experiment is repeated to get several values of W and e. The stress is found using the value of W and the area of the wire. Strain is found as $\frac{e}{l}$. A graph of stress is plotted against strain and the gradient is the Young modulus.

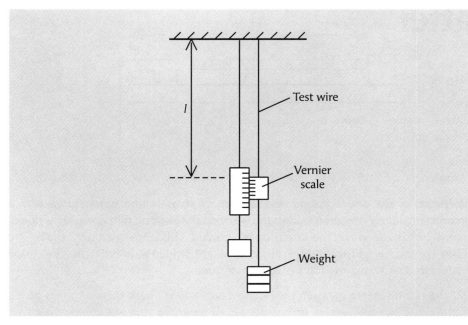

🅔 6 marks. A good diagram and the account does give the necessary detail as to how measurements are made. Marks have been lost because it is not clear how to determine stress and no attempt is made to check that the changes are elastic. It would have been better if the candidate had said that stress is plotted on the *y*-axis of the graph.

(c)

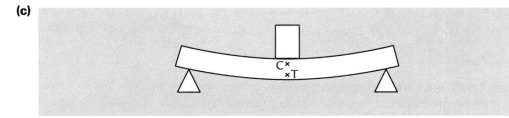

🅔 2 marks. Point C is clearly above the midline and point T is clearly below.

■ ■ ■

Answer to Question 6.1: candidate B

(a) (i) Stress = force ÷ area
(ii) Strain = extension ÷ length

🅔 1 mark. The area and the length have not been given as the cross-sectional area and the unextended length.

(b)

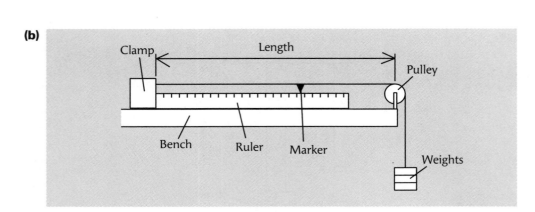

The length of the wire is found and its area of cross-section is measured with a micrometer gauge. A weight is put on the wire and the position of the marker is noted. More weight is added and the extension is recorded. This is repeated until the weight is 50 N. For each weight, the stress is found as weight divided by area. Strain is extension over length. The Young modulus is stress over strain.

2 marks. The length measured is incorrect and area of cross-section cannot be measured directly with a screw gauge. Furthermore, the means by which extension is measured is not practicable. The theory is given but no attempt is made either to plot a graph so that a check can be made for an elastic change or to obtain an average value for the modulus.

(c)

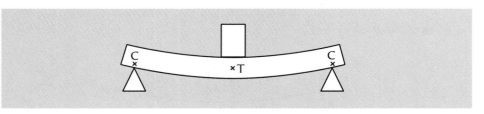

1 mark. The positions for compression are correct. However, the point T is shown on the mid-line. For a tension, it should be below.

Question 6.2

(a) Distinguish between the *elastic* and the *plastic* behaviour of a stretched material.   (3 marks)

(b) Materials are sometimes said to be ductile or to be brittle.
   (i) Identify one material that is ductile and one that is brittle.
   (ii) Sketch a graph to show the variation with stress of the strain of a material that is:
       (1) ductile
       (2) brittle   (6 marks)

(c) The Young modulus for a sample of glass is 5.5×10^{10} Pa and it has a breaking stress of 1.0×10^9 Pa. Calculate the maximum extension of a glass fibre of length 12 cm before it breaks.

(4 marks)

Total: 13 marks

■ ■ ■

Answer to Question 6.2: candidate A

(a) In an elastic material, it returns to its original shape and size. A plastic material stays deformed.

> 🄴 2 marks. The candidate has not stated that the deforming force must be removed for elastic/plastic behaviour to be observed.

(b) (i) Copper is ductile but glass is brittle.

(ii)

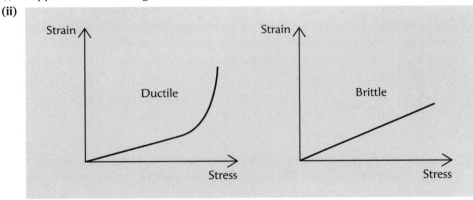

> 🄴 6 marks. In part (i) ductile and brittle materials have been identified. The graphs in part (ii) have labelled axes and the lines have the correct shape.

(c) $E = \dfrac{\text{stress}}{\text{strain}}$

$\text{strain} = \dfrac{1.0 \times 10^9}{5.5 \times 10^{10}} = 0.0182$

$\text{extension} = 0.0182 \times 12 = 0.22 \text{ cm}$

> 🄴 4 marks. The calculation is correct and the working has been explained.

■ ■ ■

Answer to Question 6.2: candidate B

(a) Elastic materials return to their original shape but plastic materials stay distorted.

> 🄴 1 mark. Original shape does not imply that there is no final change in the dimensions. The candidate has lost a second mark for not stating that the deforming force must be removed.

(b) (i) Aluminium is ductile and concrete is brittle.

(ii)

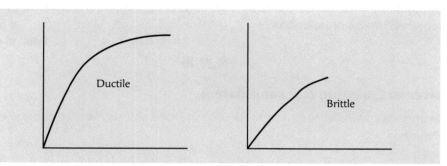

Ductile

Brittle

🄴 3 marks. In part (i) ductile and brittle materials have been identified. In part (ii) the axes have not been labelled and therefore it must be assumed that stress has been plotted on the *x*-axis. The axes for the ductile material are, therefore, incorrect. The lines have been sketched very roughly and do not show a straight-line region. Many candidates do not seem to appreciate that a sketch graph should have labelled axes and the line must be drawn carefully to represent the features of the change.

(c) Strain $= \dfrac{5.5 \times 10^{10}}{1.0 \times 10^{9}} = 55$

extension $= 55 \times 12 = 660\,\text{cm}$

🄴 2 marks. The candidate has made the fundamental error of not writing down the equation before substituting. Consequently, when making *strain* the subject of the equation, a mistake was made. Two marks have been awarded for the calculation of the extension from the strain. However, the candidate should have realised that the answer is unrealistic. It would have been worthwhile checking the working for the calculation of the strain.

Question 6.3

A student attempts to measure the change in the load on a steel wire by measuring its change in resistance. The steel has resistivity $\rho = 2.0 \times 10^{-7}\,\Omega\,\text{m}$ and Young modulus $E = 2.1 \times 10^{11}\,\text{Pa}$. The wire is 2.5 m long and has an area of cross-section of $7.9 \times 10^{-7}\,\text{m}^2$. It may be assumed that any extension of the wire does not cause a change in the area of cross-section.

(a) Calculate the extension of the wire for an additional load on the wire of 40 N.

(3 marks)

(b) Determine the change in resistance of the wire for the additional load.

(3 marks)

(c) The wire is known to extend by 3.0×10^{-5} m for every degree Celsius rise in temperature. With reference to your answers in **(a)** and **(b)**, comment on the validity of the method.

(4 marks)

Total: 10 marks

Answer to Question 6.3: candidate A

(a) $E = \dfrac{\text{stress}}{\text{strain}}$

$\text{stress} = \dfrac{F}{A} = \dfrac{40}{7.9 \times 10^{-7}} = 5.06 \times 10^7 \, \text{Pa}$

$\text{strain} = \dfrac{5.06 \times 10^7}{2.1 \times 10^{11}} = 2.41 \times 10^{-4}$

$\text{extension} = 2.41 \times 10^{-4} \times 2.5 = 6.0 \times 10^{-4} \, \text{m}$

3 marks. This is quite an involved calculation. Note that the candidate has not attempted the whole calculation in one step. Rather, it has been broken down into a number of stages, with an answer given at each of these stages. This technique is less likely to result in mistakes and, if one is made, then the examiner can see where the mistake occurs and can then award subsequent credit.

(b) Change in resistance $= \dfrac{\rho \Delta L}{A}$

$= \dfrac{2.0 \times 10^{-7} \times 6 \times 10^{-4}}{7.9 \times 10^{-7}}$

$= 1.52 \times 10^{-4} \, \Omega$

3 marks. Again, a well set-out calculation with the correct unit.

(c) This change in resistance is small and would be hard to measure. Heating the wire by 20 °C would cause the same change in length. This means that a small temperature rise would affect the reading.

3 marks. The candidate realises that this small change in resistance would be hard to measure. Also, that the expansion would be significant. The candidate did not mention that the resistivity would change with change of temperature.

Answer to Question 6.3: candidate B

(a) $E = \dfrac{\text{stress}}{\text{strain}}$

$\text{strain} = \dfrac{40}{(7.9 \times 10^{-7}) \times (2.1 \times 10^{11})} = 2.14 \times 10^{-4}$

$\text{extension} = 2.14 \times 10^{-4} \times 2.5 = 5.35 \times 10^{-4} \, \text{m}$

> 🄮 2 marks. This is quite an involved calculation. It may have been beneficial if the candidate had divided the work into a number of separate stages. The substitution is correct and so two marks have been scored. There is an error in the final answer, possibly a transcription error when reading the calculator value for the strain.

(b) Change in $R = \dfrac{2.0 \times 10^{-7} \times 5.35 \times 10^{-4}}{7.9 \times 10^{-7}}$

$= 1.35 \times 10^{-4}\,\Omega$

> 🄮 3 marks. The calculation is correct when based on the candidate's answer to (a). The candidate has not really considered the relation between the change in length and the change in resistance, but this omission is not penalised here.

(c) This change in resistance is quite small. The wire would have to be heated by 4.5 °C for the same change in length. The room temperature will not change by this amount during an experiment so temperature will not affect the result.

> 🄮 No marks. The candidate has not stated the consequence of having a small change in resistance. Furthermore, the candidate does not realise either that a small temperature change would give rise to a significant change in length or that resistivity is temperature dependent.

Gases

Question 7.1

(a) Explain what is meant by an ideal gas. (2 marks)

(b) Define the mole. (2 marks)

(c) A vessel of volume $2.5 \times 10^3 \, cm^3$ contains an ideal gas at a temperature of $27\,°C$ and a pressure of $2.0 \times 10^5 \, Pa$. Calculate the amount of gas in the vessel, given that the molar gas constant is $8.31 \, J\,K^{-1}\,mol^{-1}$. (3 marks)

(d) A further 0.05 mol of the gas is introduced into the vessel in **(c)**. The volume remains constant but the temperature rises to $40\,°C$. Calculate the new pressure of the gas in the vessel. (3 marks)

Total: 10 marks

■ ■ ■

Answer to Question 7.1: candidate A

(a) Obeys the law $pV = nRT$ at all temperatures, volumes and pressures.

> e 2 marks. The law has been stated and the conditions made clear. It would have been better if the symbols had been explained. However, the symbols used are the conventional symbols and so this omission would be overlooked.

(b) The amount of substance containing the same number of atoms/molecules as there are in 0.012 kg of carbon-12.

> e 2 marks. It would be better to refer to 'elementary entities' rather than atoms/molecules, but this is acceptable at AS/A-level.

(c) $pV = nRT$
$2.0 \times 10^5 \times 2.5 \times 10^3 \times 10^{-6} = n \times 8.31 \times 300$
$n = 0.20 \, mol$

> e 3 marks. The calculation is explained adequately. The volume has been converted to m^3 and the temperature to kelvin.

(d) $p \times 2.5 \times 10^{-3} = 0.25 \times 8.31 \times 313$
$p = 2.6 \times 10^5 \, Pa$

> e 3 marks. The candidate realises that the equation of state can be applied to the whole amount of gas in the vessel.

■ ■ ■

Answer to Question 7.1: candidate B

(a) Obeys the gas laws at all temperatures, volumes and pressures.

> e 1 mark. The laws have not been stated. It would have been better to give the universal gas equation. The conditions have been stated adequately.

(b) The amount of substance containing Avogadro's number of atoms.

> e No marks. Avogadro's number cannot be used to define the mole. Furthermore, the substance may have a molecular structure.

(c) $2.0 \times 10^5 \times 2.5 \times 10^3 \times 10^{-4} = n \times 8.31 \times 300$
$n = 20 \, \text{mol}$

> e 2 marks. The calculation has not been explained but, in this case, the substitution is clear. The volume has not been converted correctly to m^3. It is a common mistake to think that the conversion factor for cm^3 to m^3 is $\times 10^{-4}$, rather than $\times 10^{-6}$.

(d) $\frac{p}{T}$ is constant

$$\frac{p}{313} = \frac{2.0 \times 10^5}{300}$$
$$p = 2.09 \times 10^5 \, \text{Pa}$$

> e No marks. The candidate has made a fundamental error when attempting to use the pressure law equation. For this equation to apply, the mass of gas must remain constant.

Question 7.2

(a) (i) The pressure p of an ideal gas is given by the equation

$p = \frac{1}{3}nm\langle c^2 \rangle$

Explain the meaning of the symbols in this equation.

(ii) Show that the mean kinetic energy of an atom of an ideal gas is proportional to the thermodynamic (Kelvin) temperature. (7 marks)

(b) The pressure of a gas that may be assumed to be ideal is $1.5 \times 10^5 \, \text{Pa}$ at a temperature of $27 \, ^\circ\text{C}$. Given that the Boltzmann constant $k = \frac{R}{N_A}$ is equal to $1.38 \times 10^{-23} \, \text{J K}^{-1}$, determine the number of atoms in unit volume of the gas. (4 marks)

(c) The volume of one atom of the gas in **(b)** is $1.4 \times 10^{-30} \, \text{m}^3$.
(i) Calculate the actual volume of the atoms in $1.0 \, \text{m}^3$ of the gas.
(ii) Comment on your answer to (i) with reference to one of the assumptions of the kinetic theory of an ideal gas. (4 marks)

Total: 15 marks

Answer to Question 7.2: candidate A

(a) (i) n is the number of atoms per unit volume.

m is the mass of an atom.

$<c^2>$ is the mean square speed of the atoms.

(ii) Mean kinetic energy $= \frac{1}{2}m<c^2>$

$p = \frac{1}{3}nm<c^2> = \frac{2}{3}n \times \frac{1}{2}m<c^2>$ and $pV = nRT$

n, the number of atoms per unit volume $= \frac{N}{V}$

so, $\frac{2}{3}\frac{N}{V} \times \frac{1}{2}m<c^2> = \frac{nRT}{V}$

and therefore $\frac{1}{2}m<c^2> = \frac{3}{2}n\frac{RT}{N} \propto T$

> 5 marks. The symbols in part (i) have been correctly identified. In part (ii) there is some confusion in that n has been used to represent both number of atoms per unit volume and number of moles. Always use different symbols to represent different quantities. The symbol N has not been explained and the examiner is left to assume that $\frac{3}{2}n\frac{R}{N}$ is a constant.

(b) $pV = NRT$ and $k = \frac{R}{N_A}$

$pV = NkN_AT$

$p = nkT$ because $n = \frac{NN_A}{V}$

$1.5 \times 10^5 = n \times 1.38 \times 10^{-23} \times 300$

$n = 3.62 \times 10^{25}$

> 3 marks. The candidate has avoided the confusion over symbols and arrived at a correct expression. The derivation was not required. However, 1 mark has been lost because the answer, although numerically correct, has not been given with its unit. The unit is m^{-3}.

(c) (i) Volume $= 1.4 \times 10^{-30} \times 3.62 \times 10^{25}$

$= 5.1 \times 10^{-5}\,m^3$

(ii) The assumption is that the atoms have a negligible volume. In this case, $5.1 \times 10^{-5}\,m^3$ is very small.

> 2 marks. The total volume is correct. However, part (ii) is poorly answered — the assumption is that the volume of the atoms is negligible *compared with the volume of the containing vessel*. Also, the candidate should have compared the volume of $5.1 \times 10^{-5}\,m^3$ to the $1.0\,m^3$ in which the atoms are found. It is unwise to refer to a quantity as 'small', 'negligible', 'large', etc., without a comparison.

■ ■ ■

Answer to Question 7.2: candidate B

(a) (i) n is the number of atoms per unit volume.

m is the mass of an atom.

$<c^2>$ is the mean speed squared.

(ii) mean kinetic energy $= \frac{1}{2}m<c^2>$

$p = \frac{1}{3}nm<c^2> = \frac{2}{3}n \times \frac{1}{2}m<c^2>$

$pV = RT$ for one mole

so, $\frac{1}{2}m<c^2> = \frac{3}{2}\frac{RT}{N} \propto T$

e 3 marks. In part (i) $<c^2>$ has not been correctly identified. In part (ii) the special case for 1 mole of gas is being considered. The final stages in the derivation have not been explained and the examiner is left to assume that $\frac{3}{2}\frac{R}{N}$ is a constant.

(b) $p = \frac{1}{2}nkT$

$1.5 \times 10^5 = \frac{3}{2} \times n \times 1.38 \times 10^{-23} \times 27$

$n = 2.68 \times 10^{26}$

e No marks. The candidate has written down an incorrect expression for the pressure. The derivation was not required. Furthermore, Kelvin temperature has not been used and the answer has been given without its unit.

(c) (i) Volume $= 1.4 \times 10^{-30} \times 2.68 \times 10^{26}$

$= 3.75 \times 10^{-4}\,\text{m}^3$

(ii) The assumption is that the atoms have a very small volume compared with the volume of the container. In this case, $3.75 \times 10^{-4}\,\text{m}^3$ is small but we don't know the size of the container.

e 2 marks. The total volume is correct for the candidate's answer to (b). In part (ii) the assumption made in the kinetic theory is correct. It is unfortunate that the candidate has failed to realise that these atoms are in a volume of $1\,\text{m}^3$ and therefore only 0.04% (using the candidate's answer) of the volume is occupied by atoms. Note that this final comment is not strictly true; it has been assumed that the atoms can be packed tightly together without any spaces between them.

Question 7.3

(a) The first law of thermodynamics may be expressed in the form

$\Delta U = q + w$

Explain carefully the meaning of each symbol in this equation. (4 marks)

(b) An ideal gas is contained in a cylinder fitted with a frictionless piston, as illustrated below.

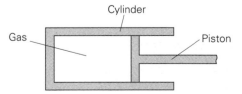

The gas is then heated. State the energy changes that occur when:

(i) the volume of the gas is held constant

(ii) the gas is allowed to expand freely against atmospheric pressure   (4 marks)

(c) When $1.0\,cm^3$ of water evaporates at its boiling point, it becomes $1600\,cm^3$ of steam at an atmospheric pressure of $1.0 \times 10^5\,Pa$.

(i) Calculate the work done against the atmosphere for the water to become steam.

(ii) The thermal energy required to evaporate this volume of water is $2.26\,kJ$. Account for the difference between this value and your answer in (i).

(5 marks)

Total: 13 marks

Answer to Question 7.3: candidate A

(a) ΔU is the increase in the internal energy.

q is the gain in thermal energy of the gas.

w is the work done on the gas.

> 🖉 4 marks. It is vital that the direction of each change is given together with the identification of the symbol. The candidate has referred to a gas. This is acceptable, although the law applies to any system.

(b) (i) No external work is done. The thermal energy increases the internal energy of the gas.

(ii) Once again the internal energy increases. However, the thermal energy is also used to do external work in pushing back the atmosphere.

> 🖉 3 marks. The distinction between the two situations has been made clearly. However, the candidate should have stated that, for an ideal gas, increase in internal energy is seen as an increase in kinetic energy of the atoms.

(c) (i) Work done $=$ pressure $\times$ volume change

$$= 1.0 \times 10^5 \times 1599 \times 10^{-6}$$

$$= 159.9\,J$$

(ii) Most of the $2.26\,kJ$ is used to increase the internal energy of the water. In this case, the energy is used to separate the molecules.

> 🖉 4 marks. In the calculation, the four significant figure answer would not be penalised, although two significant figures is appropriate. In part (ii) the candidate did not state that the kinetic energy of the water molecules would not increase.

Answer to Question 7.3: candidate B

(a) ΔU is the change in the internal energy

q is the gain in thermal energy of the gas

w is the work done on the gas

> 3 marks. All of the symbols have been identified but the direction of the change for the internal energy has not been given. This is a common error. In this context, ΔU is an *increase* in internal energy.

(b) (i) No work is done and the temperature of the gas rises.

(ii) The temperature rises and heat is required to do work to push back the atmosphere.

> 1 mark. The candidate was asked to give the energy changes, not the temperature changes. The external work done in part (ii) has been mentioned.

(c) (i) Work done $= 1.0 \times 10^5 \times 1599 \times 10^{-4}$

$= 15\,990\,J$

(ii) The rest of the 2.26 kJ is used to increase the internal energy of the water.

> 2 marks. In the calculation, it is necessary to assume that the candidate is using the equation $\Delta W = p\Delta V$. This is fairly obvious, but there is an error when converting cm^3 to m^3 (a very common error!). It is always advisable to give the equation in symbols before any substitution is made. In part (ii) the candidate has not distinguished between the different components of the internal energy.

Waves and oscillations

Question 8.1

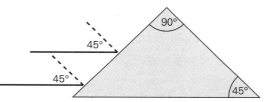

(a) Distinguish between the *refraction* and the *diffraction* of light. (4 marks)

(b) With the aid of labelled diagrams, explain what is meant by *critical angle*.
(4 marks)

(c) The diagram below shows two rays of light incident on one face of a right-angled prism at an angle of incidence of 45°.

The refractive index of the glass of the prism when the prism is in air is 1.52.
(i) Calculate the critical angle for light in the prism.
(ii) Copy the diagram and on your copy trace the paths of the two rays through the prism and back into the air. (6 marks)

Total: 14 marks

■ ■ ■

Answer to Question 8.1: candidate A

(a) Refraction is the bending of light when it moves from one medium to another. In diffraction, the light changes direction when it passes by an edge or a slit.

> e 3 marks. It would be better to refer to a change in the direction of a wave, rather than 'bending of light'. The candidate has not mentioned that refraction occurs as a result of a change in wave speed.

(b)

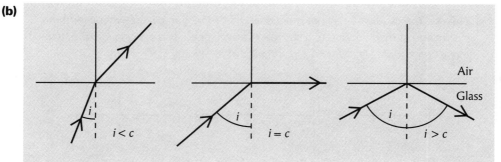

When the ray approaches the boundary from the more dense medium, it bends away from the normal. If the ray is incident at the critical angle, the ray emerges along the surface. For angles greater than this, all the light is reflected. At the critical angle, $\sin c = \dfrac{1}{n}$.

> 4 marks. Although not essential for this answer, there is always some reflection at the surface. If the question had asked for an explanation of total internal reflection, then this partial reflection would have been important.

(c) (i) $\sin c = \dfrac{1}{1.52}$

$c = 41°$

(ii) $n = \dfrac{\sin i}{\sin r}$

$1.52 = \dfrac{\sin 45}{\sin r}$

$r = 28°$

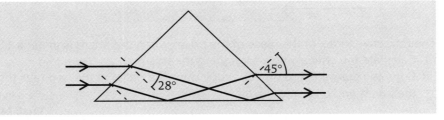

> 6 marks. Angles c and r have been calculated correctly and the ray trace is satis-factory. It is not necessary to measure the angles carefully. If the size of each angle is marked, then it is sufficient for the angles to look about right.

■ ■ ■

Answer to Question 8.1: candidate B

(a) Refraction is bending of light when it enters a medium. Diffraction is the bending of light as it passes through a slit.

> 2 marks. The candidate has not mentioned that refraction occurs whenever there is a change of medium due to a change in wave speed. The explanation of diffrac-tion is limited to a slit, without any mention of an edge.

(b)

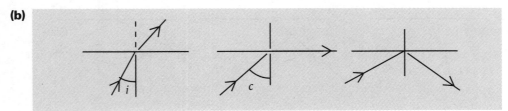

For angles of incidence greater than the critical angle, the ray is not refracted but is totally internally reflected.

🄔 2 marks. The basic diagrams are just about satisfactory but some detail is missing. The ray must be incident on the boundary from the more dense medium. Also, the expression '$\sin c = \frac{1}{n}$' has not been included. For total internal reflection, the angles of incidence and reflection do not appear to be equal.

(c) (i) $\sin c = \dfrac{1}{1.52}$

$c = 41°$

(ii)

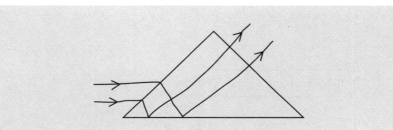

🄔 3 marks. The calculation in part (i) is correct. However, the diagram in part (ii) is very poor. *Always use a ruler when drawing straight lines*. In this case, the rays do not cross the normal on entering the prism. It is always wise to draw in the normal. Total internal reflection has been shown at the base of the prism but the diagram has been so poorly drawn that it appears as if the rays are normal to the third face. Candidates lose many marks through a failure to take a reasonable amount of care with ray diagrams — use a ruler and draw the angles at about the correct size.

Question 8.2

(a) (i) State what is meant by the *frequency, wavelength* and *speed* of a progressive sound wave.
(ii) Write down the equation relating frequency *f*, wavelength λ and speed *v* for a wave. (4 marks)

(b) Outline an experiment to determine the frequency of a sound wave using a cathode-ray oscilloscope (c.r.o.). (6 marks)

Total: 10 marks

Answer to Question 8.2: candidate A

(a) (i) Frequency is the number of complete oscillations made per unit time by a particle in the wave.
Wavelength is the distance between two points vibrating in phase.
Speed is the distance moved per unit time by a wavefront in the wave.

(ii) $v = f\lambda$

e 3 marks. In part (i) the definitions of frequency and speed are satisfactory, with the ratios being made clear. The wave in this question is progressive and so the definition of wavelength would have been correct if the candidate had stated that the points must be neighbouring points. Part (ii) is correct.

(b)

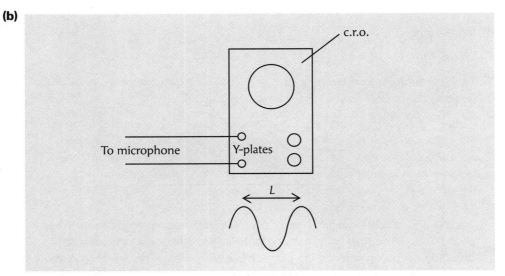

The microphone is connected to the Y-plates of the c.r.o. With the sound switched on, the length L of a wave is measured. The time-base T (in $s\,cm^{-1}$) is read from the dial. The time for one wavelength is TL and so the frequency is $\frac{1}{TL}$.

e 4 marks. When giving the method, the candidate has not mentioned the adjustment of the controls. The Y-plate sensitivity should be adjusted to give a wave with a 'large' amplitude and the time-base should be chosen so that there is at least one complete wave on the screen. Adequate theory has been included.

■ ■ ■

Answer to Question 8.2: candidate B

(a) **(i)** Frequency is the number of vibrations in one second.
Wavelength is the distance moved by the wave during one vibration of the source.
Speed is the speed at which the energy of the wave travels.
(ii) $v = f\lambda$

e 2 marks. In part (i) the definition of frequency fails to include the ratio and also introduces a unit (second) into a definition. Rather than 'in one second', the candidate should have written 'per unit time'. The definition of wavelength is satisfactory. However, the quantity *speed* is not defined. Part (ii) is correct.

(b)

With the microphone connected to the c.r.o., the length of one wave is measured. This length is multiplied by the time-base to give the frequency.

> 🖰 1 mark. This answer is typical of a candidate who does not write in sufficient detail and does not include relevant points on diagrams. It is not clear what connections are made between the microphone and the c.r.o. The adjustments to the c.r.o. have not been mentioned and it is not clear what measurements are made — a sketch of the trace seen on the screen would have been helpful. The theory is faulty in that the periodic time has been found, not the frequency.

Question 8.3

(a) State what is meant by the *amplitude* of an oscillation.   (1 mark)

(b) (i) Define simple harmonic motion.
      (ii) Sketch a graph to show the variation with displacement x of the
          acceleration a of a particle undergoing simple harmonic motion.   (5 marks)

(c) Some grains of sand rest on a horizontal plate. The plate vibrates vertically
      with simple harmonic motion of amplitude 1.5 mm.
      (i) Calculate the frequency of vibration of the plate when its maximum
          acceleration is equal to the acceleration of free fall ($9.8\,\mathrm{m\,s^{-2}}$).
      (ii) Suggest what will happen to the grains of sand when the frequency
          of vibration is increased beyond that calculated in (i).   (6 marks)

Total: 12 marks

■ ■ ■

Answer to Question 8.3: candidate A

(a) It is the maximum distance from the equilibrium position.

> 🖰 1 mark. Note that amplitude should not be defined in terms of maximum displacement because amplitude is a scalar and displacement is a vector.

(b) (i) The motion of a particle such that the force acting on it is proportional to its
          distance from a fixed point and is directed towards that point.

(ii)

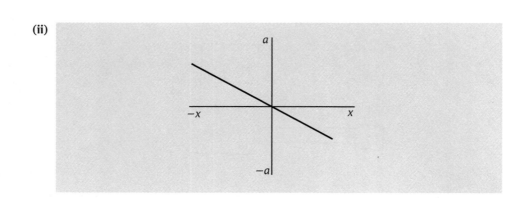

🄴 4 marks (2 marks for each part). In part (i) it would be have been better to refer to acceleration, rather than force. However, force is acceptable here because at AS/A-level, the mass of the particle may be assumed to be constant. In part (ii) the graph quite correctly shows $a \propto -x$, but the maximum displacements in the $+x$ and in the $-x$ directions are not equal.

(c) (i) $a = (2\pi f)^2 x$
$9.8 = (2\pi f)^2 \times 1.5 \times 10^{-3}$
$f = 13\,Hz$

(ii) The maximum acceleration will be greater than g and so the sand will jump off the plate as the plate begins to move downwards.

🄴 6 marks. The calculation in part (i) is successful and presented with adequate explanation. It is correct to give the answer to two significant figures (the data are to two figures) but the answer of 12.9 Hz would normally be accepted. Part (ii) is also answered well: the observation is fully described and explained.

■ ■ ■

Answer to Question 8.3: candidate B

(a) It is the maximum distance moved from the rest position.

🄴 No mark. Maximum distance moved increases with the number of oscillations!

(b) (i) A particle vibrates so that its acceleration is proportional to its distance from a fixed point.

(ii)

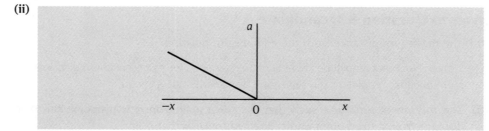

e 2 marks. 1 mark is generously given for part (i) — in fact, this particle would not vibrate! The candidate has failed to state that the acceleration is directed towards the fixed point. In part (ii) the graph shows $a \propto -x$ but in one direction only. Pay careful attention to the wording of the question. Many candidates would automatically draw a sinusoidal wave.

(c) **(i)** $a = (2\pi f)^2 x$
$9.8 = (2\pi f)^2 \times 1.5$
$f = 0.41\,\mathrm{s^{-1}}$
(ii) The acceleration is larger so the sand falls off the plate.

e 2 marks. In part (i) the candidate failed to use amplitude measured in metres. Otherwise the calculation is correct with the correct alternative unit. No marks are given for part (ii). The candidate has not referred to the maximum acceleration, and larger than what? The sand does not fall off the plate but, rather, it leaves the surface of the plate whenever the acceleration of the plate is greater than, and in the same direction as, g.

Question 8.4

(a) The graph below shows the variation with time of the displacement of two waves, A and B.

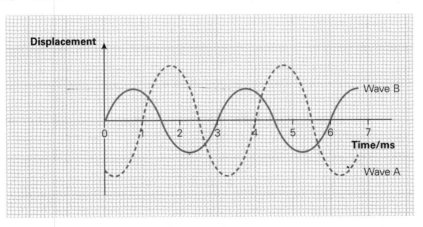

Determine:
(i) the frequency of the waves
(ii) the phase angle with which wave A leads wave B   (4 marks)

(b) (i) Explain what is meant by the *superposition* of two waves.
(ii) Two waves X and Y have the same amplitude A and intensity I. Wave X leads wave Y by a phase angle of 40°. At one time, the displacement of wave Y is $+A$ at a particular point. Assuming that wave X passes through the same point, determine:

(1) the displacement of wave X at this point at the same time
(2) the resultant displacement at the point
(3) the resultant intensity in terms of *I*

(6 marks)

Total: 10 marks

■ ■ ■

Answer to Question 8.4: candidate A

(a) (i) Period of waves = 3.0 ms

Frequency = $\frac{1}{T}$ = 333 Hz

(ii) Wave A is one square ahead of wave B and one period is three squares.
Phase angle = $2\pi \times \frac{1}{3} = \frac{2}{3}\pi$ rad

e 4 marks. Note that explanation has been given so that, should there be an arithmetical error, some marks could still be scored.

(b) (i) When two waves meet, the resultant displacement is the sum of the individual displacements.

(ii) (1) Displacement = $A\sin40$ = +0.64A
(2) Resultant displacement = +1.64A
(3) Final intensity $\propto (1.64A)^2 \propto 2.7A^2$
but intensity $I \propto A^2$
final intensity = 2.7*I*

e 4 marks. Part (i) has been explained correctly in terms of displacements, not amplitudes. It is a common mistake to try to discuss amplitudes. The working for part (ii) is incorrect. Even the best candidates make this mistake. If the displacement is *A*, then a lead of 40° will give a displacement of $A\cos40$. The answer should be 0.77A. The remaining parts of the question are correct when based on the candidate's answer to part (ii) (1) and are given full subsequent credit. The correct answers are 1.77A and 3.1*I*.

■ ■ ■

Answer to Question 8.4: candidate B

(a) (i) Period of waves = 3.0 ms

Frequency = $\frac{1}{T}$ = 3.33 × 10⁵ Hz

(ii) Wave B is one-third of a period ahead.
Phase angle = $360 \times \frac{1}{3}$ = 120°

e 3 marks. A common error has been made in part (i) — confusion of μ and m. Fortunately, some working has been shown so that one of the two available marks could be awarded. Part (ii) is correct.

AS/A-Level Physics

(b) (i) The resultant amplitude of two waves is the sum of the individual amplitudes.

(ii) (1) Displacement $= A\sin40 = +0.64A$
(2) Resultant displacement $= +1.64A$
(3) Final intensity $= 1.64I$

e 1 mark. Part (i) is wrong. The candidate has not mentioned that waves must meet and the discussion is in terms of amplitude, not displacement. The working for part (ii) is incorrect. The displacement should be $A\cos40 = +0.77A$. Part (ii) (2) is correct when based on the candidate's answer to part (ii) (1). No marks are awarded in part (ii) (3). The candidate does not appreciate that intensity $\propto$ amplitude^2.

Question 8.5

(a) The following graph shows the variation with time of the velocity of a particle undergoing simple harmonic motion.

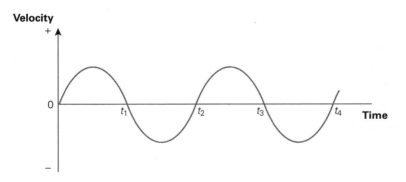

Using the same values of time, plot graphs, one in each case, to show the variation with time of:
(i) the displacement of the particle
(ii) the acceleration of the particle   (4 marks)

(b) The displacement x, measured in millimetres, of an oscillating particle is given by the equation

$x = 1.5\sin2200t$

where t is in seconds.
(i) Give a quantitative description of the motion.
(ii) The particle has a mass of 5.0 g. Calculate the maximum force acting on the particle due to its oscillating motion.   (7 marks)
Total: 11 marks

$F = m \times a$ ■ ■ ■

Answer to Question 8.5: candidate A

(a) (i)

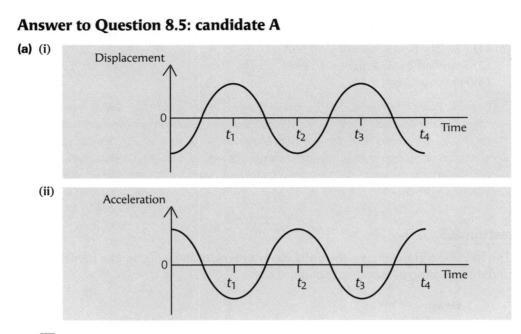

(ii)

 4 marks. Careful sketching has paid dividends! The phases and times are correct and the waves drawn are approximately sinusoidal.

(b) (i) The amplitude is 1.5 mm.
$2 \times \pi \times f = 2200$
frequency $= 350$ Hz
(ii) Acceleration $= -(2\pi f)^2 x$
force $= ma = 0.005 \times 2200^2 \times 1.5$
$= 36\,300$ N

 5 marks. In part (i) the candidate has not stated that, because the motion is sinusoidal, it must be simple harmonic. In part (ii) the calculation is explained well, but the amplitude was left in mm. The correct answer is 36.3 N.

■ ■ ■

Answer to Question 8.5: candidate B

(a) (i)

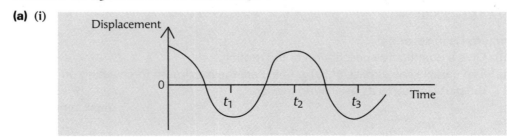

(ii)

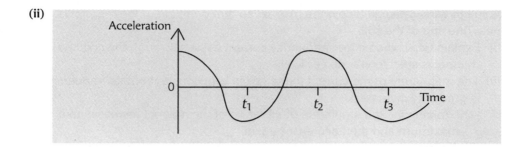

📝 2 marks. In part (i) the phase for the displacement/time graph is wrong. Also, the poor quality of the sketch shows differences in the positive and negative displacements. The sketch graph in part (ii) for acceleration/time is barely acceptable.

(b) (i) The amplitude is 1.5 mm.
Angular frequency = 2200 Hz
(ii) Acceleration = $-\omega^2 x = 7.26 \times 10^6$
force = $ma = 0.005 \times 7.26 \times 10^6$
$= 36\,300\,\text{N}$

📝 3 marks. In part (i) the candidate has not stated that the motion is simple harmonic. Also, the unit for angular frequency is incorrect. In part (ii) the calculation is explained adequately, but the amplitude has been left in mm, rather than converted to metres. The correct answer is 36.3 N.

Question 8.6

(a) (i) Explain what is meant by the *damping* of the oscillations of a vibrating body.
(ii) Sketch a graph to show the variation with time of the displacement of a particle undergoing simple harmonic motion with light damping.   (5 marks)

(b) A magnet is suspended from a spring as shown below.

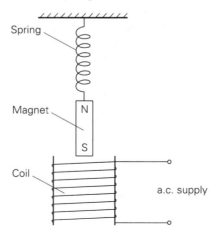

A coil of wire is placed below the magnet so that one pole of the magnet is near one end of the coil.
  (i) Explain why, when a low-frequency current passes through the coil, the magnet is seen to vibrate vertically.
  (ii) The magnitude of the current in the coil is kept constant but its frequency is gradually increased.
     (1) Explain why the amplitude of vibration of the magnet increases to a maximum and then decreases again.
     (2) A piece of light card is attached to the magnet with its plane normal to the axis of the magnet. Describe how the amplitude of vibration of the magnet will be affected as the frequency of the current is increased from a low value.   (9 marks)

Total: 14 marks

Answer to Question 8.6: candidate A

(a) **(i)** The amplitude of vibration decreases due to resistive forces dissipating the energy of the object.

  (ii)

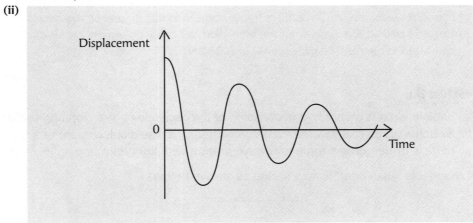

  ℮ 4 marks. In part (i) it was not stated that the amplitude decreases progressively (or better still, exponentially) with time. The graph in part (ii) does show a progressive decrease in amplitude with the period remaining constant.

(b) **(i)** The coil acts as a magnet and attracts the pole of the magnet. When the current reverses, the magnetic field in the coil changes and repels the magnet, making it vibrate.

  (ii) **(1)** This is resonance. The magnet is made to vibrate. When this forcing frequency is equal to the natural frequency of vibration of the magnet, the magnet will vibrate with a maximum amplitude.

(2)

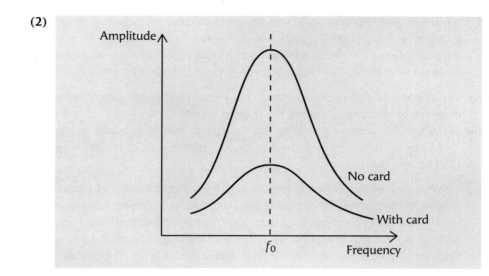

The amplitude is always less and the peak is less sharp.

 8 marks. In part (i) it is assumed that attraction occurs first. This is acceptable at AS/A-level. However, the magnetic field was said to change, not reverse, when the current reverses. All the points have been covered in part (ii). Note that a sketch diagram has been drawn for part (ii) (2). Without that, more explanation would have been necessary. In particular, no mention has been made of the position of the peak.

■ ■ ■

Answer to Question 8.6: candidate B

(a) (i) The amplitude of vibration decreases due to forces acting on the object.

(ii)

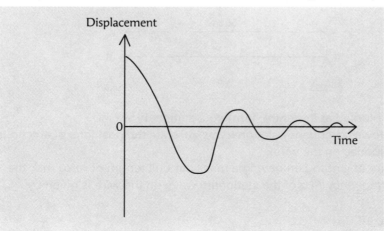

🅔 1 mark. In part (i) it was not stated that the amplitude decreases progressively (or better still, exponentially) with time. The nature of the force and its effect on the motion were not made clear. The graph in part (ii) does show a progressive decrease in amplitude but with a decreasing period. This is a common fault.

(b) (i) The coil acts as a magnet and attracts the magnet. When the current changes, the magnetic field changes, making the magnet vibrate.

(ii) (1) When the frequency forcing the magnet to vibrate is equal to the natural frequency of the magnet, the magnet will vibrate with its biggest amplitude.

(2) The amplitude is less.

🅔 3 marks. In part (i) a common mistake has been made: instead of considering reversal of the current, the candidate merely mentions 'changing' current. In part (ii) no mention has been made of resonance. The idea that impressed frequency is equal to the natural frequency of vibration of the magnet is poorly expressed. Note that a sketch diagram has not been drawn in part (ii) (2). The description is scant. Marks could have been scored for a good sketch.

Question 8.7

(a) Distinguish between a *progressive* wave and a *stationary* wave, making reference to:
(i) transfer of energy
(ii) amplitude of oscillation of particles
(iii) phase of oscillating particles   (7 marks)

(b) A metal wire has a length of 150 cm and mass 9.3 g. A section of this wire is stretched between two points A and B, separated by a distance of 58 cm, as shown in the diagram below.

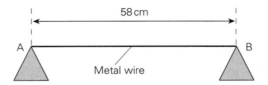

The wire is plucked at its centre, forming a stationary wave.
(i) For the lowest frequency stationary wave, state the positions of any nodes and antinodes on the wire.
(ii) For a wire of length L/m having a mass per unit length of m/kg m^{-1}, the lowest frequency f/Hz of the stationary wave on the wire is given by

$$f = \frac{1}{2L}\sqrt{\frac{T}{m}}$$

where T/N is the tension in the wire. Calculate the tension in the wire for a frequency of 76 Hz.

(iii) Without changing the length or tension of the wire, suggest how the wire could be made to vibrate so that a frequency of 152 Hz predominates.

(7 marks)

Total: 14 marks

Answer to Question 8.7: candidate A

(a) (i) Energy moves along on a progressive wave. In a stationary wave, energy is not transferred.

(ii) In a progressive wave, all particles have the same amplitude. The amplitude of neighbouring particles is different in a stationary wave.

(iii) In a progressive wave, particles vibrate out of phase with one another. In a stationary wave, neighbouring particles vibrate in phase.

⊖ 4 marks. Further detail was required to score full marks. For example, in part (i) it could have been stated that the energy is stored in a stationary wave as kinetic and potential energy of the vibrating particles. In part (ii) mention could have been made of nodes and antinodes. In part (iii) the fact that all particles in an internodal loop vibrate in phase and in antiphase with those in neighbouring loops could have been discussed. The fact that particles one wavelength apart in a progressive wave vibrate in phase could also have been mentioned. In such an answer, not all the points would be necessary for full credit. However, candidates should look carefully at the mark allocation and, having seen that seven marks are available, should realise that the question would require some detail of the situations.

(b) (i) A node at each end and an antinode at the centre.

(ii) $m = \dfrac{9.3 \times 10^{-3}}{1.5} = 6.2 \times 10^{-3} \text{ kg m}^{-1}$

$76 = \dfrac{1}{2 \times 0.58} \times \sqrt{\dfrac{T}{6.2 \times 10^{-3}}}$

$T = 48.2 \text{ N}$

(iii) The wire should be plucked 14.5 cm from one end.

⊖ 6 marks. Part (i) has been answered correctly, although it could have been completed just as successfully by means of a sketch. The calculation is correct. The formula had been given and therefore the main point is to take care with units. In part (iii) it is correct to pluck the wire at an antinode but also the wire must be touched at a node, i.e. the wire should be touched lightly at its centre.

Answer to Question 8.7: candidate B

(a) **(i)** Energy travels along on a progressive wave.

   (ii) Amplitude is different in a stationary wave but is the same in a progressive wave.

   (iii) In a progressive wave, particles don't vibrate in phase, but in stationary waves, they do.

> 🅴 3 marks. Detail is lacking (see comments for candidate A). A common omission has been made in part (i): because there is no energy transfer for a stationary wave, this has not been stated. It is sometimes as important to state what does not happen as it is to state what does! In parts (ii) and (iii) ideas are poorly expressed and detail is lacking.

(b) **(i)** There are nodes at the ends and an antinode in the middle.

   (ii) $m = \dfrac{9.3}{1.5} = 6.2 \times 10^{-3}$

   $76 = \dfrac{1}{116} \times \sqrt{\dfrac{T}{6.2 \times 10^{-3}}}$

   $T = 4.82 \times 10^5 \, \text{N}$

   (iii) The wire could be bowed, rather than plucked.

> 🅴 3 marks. Part (i) has been answered correctly. The calculation is not correct in that the length was not converted from centimetres to metres. The candidate should have realised there was an error because the tension is far too large. Part (iii) is incorrect. The first overtone must be made to predominate.

Question 8.8

(a) Explain what is meant by *coherent* sources. (1 mark)

(b) Describe an experiment to demonstrate two-source interference for light. Include in your account suitable dimensions for the apparatus used. (5 marks)

(c) The interference pattern produced by two coherent sources of light S_1 and S_2 is viewed on a screen. Initially the sources are of equal intensity. State, with reasons, the effect on the interference pattern when the following changes are made, in turn, to the sources S_1 and S_2:

   (i) the wavelength of the light from the coherent sources is increased

   (ii) the separation of the sources is decreased

   (iii) the intensity of light from S_1 is increased but that from S_2 is unchanged

   (7 marks)

Total: 13 marks

■ ■ ■

Answer to Question 8.8: candidate A

(a) The sources must have a constant phase difference.

> 🅔 1 mark. This is correct. Constant phase difference implies constant frequency so there is no need to mention it.

(b)

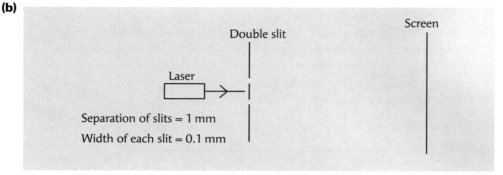

The apparatus is set up as shown. The room should not be well-lit. Light and dark fringes are seen on the screen.

> 🅔 4 marks. There is little need for a detailed description because the diagram makes everything clear. In such questions it is worthwhile making sure that the diagram is accurate. The candidate has failed to indicate the distance between the double slit and the screen. It is implied that a laser emits monochromatic light, but it would have been advisable to indicate this.

(c) (i) The separation of the fringes increases because $\lambda = \dfrac{ax}{D}$ and a and D are constant.

(ii) The separation of the fringes increases because $\lambda = \dfrac{ax}{D}$ and λ and D are constant.

(iii) The bright fringes are brighter because the amplitudes are added at a maximum. The dark fringes remain dark.

> 🅔 3 marks. Questions such as this are frequently answered poorly. In parts (i) and (ii) the answers are correct but would have been better if the symbols had been explained. The candidate has not stated that the bright fringes would remain bright and the dark fringes dark because the amplitudes have not changed. In part (iii) the dark fringes would, in fact, be lighter because the two waves no longer completely cancel one another. No mention has been made of fringe separation.

■ ■ ■

Answer to Question 8.8: candidate B

(a) The frequencies must be the same.

No mark. The frequencies may be the same but if the waves are not continuous, then the phase difference will not be constant.

(b)

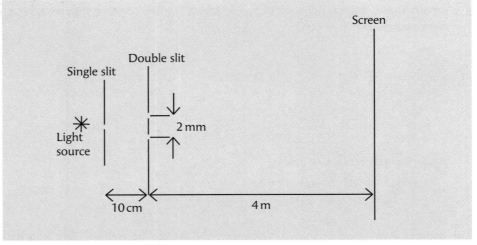

The room must be darkened. Fringes are seen on the screen.

 3 marks. It has not been stated that the light source must be monochromatic (either that or a colour filter should have been used). The distance between the slits of the double slit is acceptable, but rather large. The width of each slit has not been given.

(c) **(i)** The separation increases because $\lambda = \frac{ax}{D}$.

(ii) The separation decreases because the formula $\lambda = \frac{ax}{D}$ still applies.

(iii) The fringes are brighter because the amplitudes add up and one of them is bigger.

 2 marks. The brightness of the light and dark fringes has not been considered in parts (i) and (ii). Also, in part (ii) a common mistake has been made: the separation does, in fact, increase. This error is probably caused by not making the separation x the subject of the formula $\lambda = \frac{ax}{D}$. Another common fault occurs in part (iii): referring to 'fringes' does not distinguish between light fringes and dark fringes.

Question 8.9

(a) Explain briefly how diffraction of waves may be demonstrated using a ripple tank. (4 marks)

(b) The diagram below illustrates plane waves incident on a narrow slit and on a wide slit.

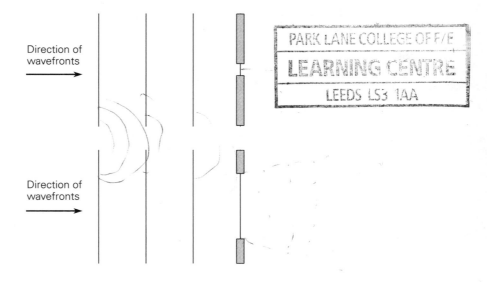

Copy and complete the diagram to show three wavefronts that have emerged from each slit. (4 marks)

(c) A diffraction grating has 2500 lines per centimetre. Light of two wavelengths, 650 nm and 700 nm, is incident normally on the grating. Determine:
(i) the numbers of complete orders observed on each side of the zero order
(ii) the angle between the light of the two wavelengths in the highest order
(5 marks)

Total: 13 marks

Answer to Question 8.9: candidate A

(a)

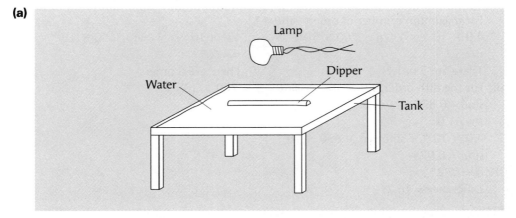

The glass-bottomed tank is set up as in the diagram. The ripples are made using a long straight dipper that is vibrating. The waves are seen on white paper below the tank.

e 4 marks. The apparatus and the means by which the ripples are observed has been outlined.

(b)

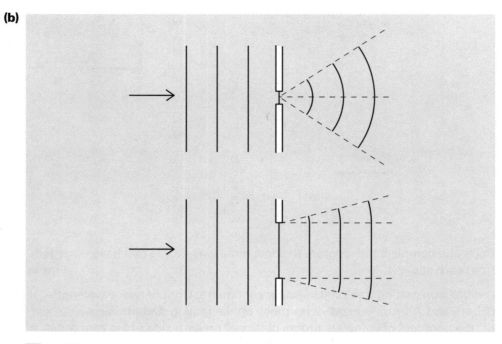

e 4 marks. The candidate has taken some care over the diagrams and consequently has scored full marks. The wavelength is constant. Circular arcs are produced at the narrow slit and plane waves with curved ends are produced at the wide slit.

(c) (i) $d\sin\theta = n\lambda$, where $d = \dfrac{1}{2500}$ cm $= 4.0 \times 10^{-4}$ cm

For maximum number of orders, $\sin\theta = 1$

$4.0 \times 10^{-6} \times 1 = n \times 700 \times 10^{-9}$ $4.0 \times 10^{-6} \times 1 = n \times 650 \times 10^{-9}$

$n = 5.7$ $n = 6.2$

There are 5 orders   There are 6 orders

(ii) For the fifth order, $4.0 \times 10^{-6} \times \sin\theta = 5 \times 700 \times 10^{-9}$

$\sin\theta = 0.875$

$\theta = 61.0°$

$4.0 \times 10^{-6} \times \sin\alpha = 6 \times 650 \times 10^{-9}$

$\sin\alpha = 0.975$

$\alpha = 77.2°$

Difference $= 16.2°$

e 5 marks. The calculations are correct, with adequate explanation.

■ ■ ■

Answer to Question 8.9: candidate B

(a)

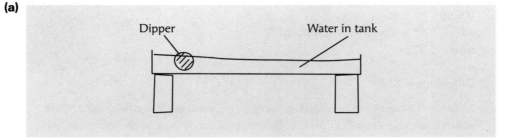

The dipper vibrates in the water producing ripples that are seen as shadows moving across the floor.

> ℮ 3 marks — rather generously. Most of the marks have been earned from the diagram. A means of illumination has been omitted.

(b)

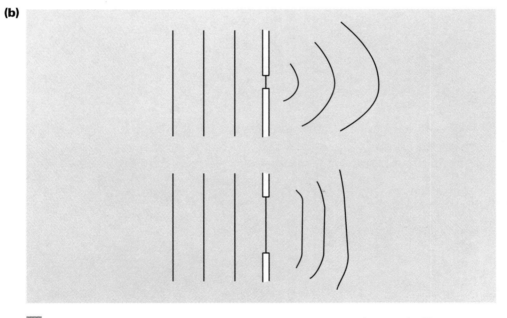

> ℮ 1 mark. Carelessly sketched diagrams like these score very few marks. The wavelength is not constant. For the narrow slit, the arcs are not circular and are not centred on the narrow slit. The shape of the waves for the wide slit is acceptable.

(c) (i) $d = \dfrac{1}{250\,000} = 4.0 \times 10^{-6}\,\text{m}$

When $\sin\theta = 1$

$4.0 \times 10^{-6} \times 1 = n \times 700 \times 10^{-9}$

$n = 5.7$

There are 6 orders

(ii) For the sixth order, $4.0 \times 10^{-6} \times \sin\theta = 6 \times 700 \times 10^{-9}$
$\sin\theta = 1.05$
$\theta = 90°$
$4.0 \times 10^{-6} \times \sin\alpha = 6 \times 650 \times 10^{-9}$
$\sin\alpha = 0.975$
$\alpha = 77.2°$
Difference $= 12.8°$

e 2 marks. It would have been better had the candidate explained why the 700 nm wavelength was chosen in part (i). The mistake made here is to 'round up' the value of n. In part (ii) the calculation based on $n = 6$ is correct for the 650 nm wavelength. The value of $\sin\theta$ for the 700 nm wavelength is not possible.

Circular motion

Question 9.1

(a) Define the radian. (2 marks)

(b) A stone is attached to a string. The stone rotates in a circle, centre C and radius 79 cm, at constant speed in a vertical plane, as shown in the diagram below.

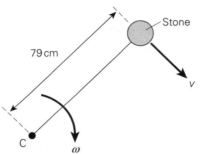

The stone has mass 45 g and it takes 0.90 s to complete one revolution. Determine, for this stone:

(i) its linear speed v
(ii) its angular speed ω
(iii) its centripetal acceleration (6 marks)

(c) Calculate the tension in the string when the string is:

(i) horizontal
(ii) vertical with the stone below C
(iii) vertical with the stone above C

The acceleration of free fall = 9.8 m s^{-2}. (5 marks)

(d) Suggest why, in practice, it would be difficult to maintain a constant angular speed of the stone. (3 marks)

Total: 16 marks

■ ■ ■

Answer to Question 9.1: candidate A

(a) Angle at the centre of a circle produced by an arc equal in length to the radius.

> 2 marks. It would have been more precise if the candidate had referred to the angle subtended at the centre of the circle.

(b) (i) $v = \dfrac{2\pi r}{T}$

$$\dfrac{2 \times \pi \times 0.79}{0.90} = 5.51\,\text{m}\,\text{s}^{-1}$$

(ii) $\omega = \dfrac{2\pi}{T}$

$$\dfrac{2 \times \pi}{0.90} = 6.98\,\text{rad}\,\text{s}^{-1}$$

(iii) Acceleration $= \omega^2 r = 38.5\,\text{m}\,\text{s}^{-2}$

 6 marks. The answers are correct and adequate explanation has been given.

(c) Centripetal force $= mr\omega^2 = 0.045 \times 0.79 \times 6.98^2 = 1.73\,\text{N}$
weight $= 0.045 \times 9.8 = 0.44\,\text{N}$
(i) When horizontal, tension $= 1.73\,\text{N}$
(ii) When stone below C, tension $= 1.73 + 0.44 = 2.17\,\text{N}$
(iii) When stone above C, tension $= 1.73 - 0.44 = 1.29\,\text{N}$

 5 marks — a good answer. Remember to calculate separately the weight and the centripetal force before attempting to find the tension in the string.

(d) The stone slows down when rising and speeds up when going down. Energy would have to be put in or taken out of the stone.

 1 mark. The candidate recognises that there is a need to compensate for changes in speed due to the stone rising or falling. However, the difficulty of doing this has not been discussed. A tangential force would be required to maintain constant speed and this could not be supplied by the string. A rigid rod would be required.

■ ■ ■

Answer to Question 9.1: candidate B

(a) There are 2π radians in one revolution ($360°$).

 No marks. The statement given is not a definition of the radian.

(b) (i) $v = \dfrac{2\pi r}{T}$

$$= 2\pi \times \dfrac{79}{0.90} = 551\,\text{cm}\,\text{s}^{-1}$$

(ii) $\omega = \dfrac{2\pi}{0.90} = 6.98\,\text{s}^{-1}$

(iii) Acceleration $= 43.8\,\text{m}\,\text{s}^{-2}$

@ 3 marks. The answer to (i) is correct. In part (ii) one mark has been lost because the wrong unit has been given. There are no marks for part (iii) — there is no explanation and the answer is wrong. Always write down the formulae you are using; marks may well be given for these.

(c) (i) When horizontal, tension $= ma = 0.045 \times 43.8 = 1.97\,N$
(ii) When stone below C, tension $= 1.97\,N$
(iii) When stone above C, tension $= 1.97\,N$

@ 2 marks. Both of the available marks for part (i) have been given. Although the correct answer is $1.73\,N$, the answer of $1.97\,N$ is correct when the candidate's value of acceleration is used. There are no marks for (ii) or (iii) because the candidate did not understand that the weight of the stone would either be in addition to the centripetal force or contribute towards it.

(d) The stone changes its speed as it goes round.

@ No marks. The candidate has done little more than repeat the question. The candidate should look at the mark allocation to realise that a one-sentence answer that gives little or no further information could not possibly be sufficient to score three marks.

Momentum

Question 10.1

(a) (i) State Newton's second law of motion.

(ii) On the basis of this law, explain why, for an object of constant mass m, its acceleration a is related to the resultant force F acting on it by the expression

$$F = ma$$

(5 marks)

(b) A toy rocket contains water and air at high pressure. Initially, the rocket and its contents have a total mass of 650 g. When the rocket is fired, the constant mass of water ejected from the rocket each second is 360 g with a speed of $9.6 \, \text{m s}^{-1}$. Calculate:

(i) the change in momentum per second of the water as it is ejected

(ii) the time elapsed between firing the rocket and the rocket lifting off the ground. The acceleration of free fall is $9.8 \, \text{m s}^{-2}$.

(5 marks)

(c) Give an explanation as to why the acceleration of the rocket as it ascends is not constant.

(3 marks)

Total: 13 marks

Answer to Question 10.1: candidate A

(a) (i) The force acting on the object is proportional to its rate of change of momentum in the direction of the force.

(ii) Momentum $= mv$

so, $F = \dfrac{(MV - mv)}{t}$

but, $M = m$

and $F = \dfrac{m(V - v)}{t}$

Now, $a = \dfrac{(V - v)}{t}$

$\therefore F = ma$

> e 5 marks. This is a good answer. The candidate has included direction in part (i) — this is often omitted. In part (ii) acceleration has been defined and it has been made clear how constant mass affects the formula.

(b) (i) Change = mass per second × change in speed

$= 0.36 \times 9.6 = 3.46 \, \text{kg m s}^{-1}$

(ii) Force $= 3.46 \, \text{N} =$ weight of rocket on take off

$$\text{Mass to be lost} = 0.65 - \left(\frac{3.46}{9.8}\right)$$
$$= 0.297\,\text{kg}$$
$$\text{Time} = \frac{297}{360} = 0.83\,\text{s}$$

 5 marks. The calculation is correct and explained adequately.

(c) The force is constant and the mass is decreasing. Since $F = ma$, the acceleration must increase.

 3 marks. Each factor (force, mass and acceleration) has been considered and the outcome has been made clear.

■ ■ ■

Answer to Question 10.1: candidate B

(a) (i) Force is equal to change in momentum per second.

(ii) Force $= \dfrac{(mv - mu)}{t}$

$\qquad\quad = \dfrac{m(v - u)}{t}$

$a = \dfrac{(v - u)}{t}$

so $F = ma$

 2 marks. No marks have been scored for the definition. Force being equal to rate of change of momentum is only true when units have been defined. Furthermore, the candidate has included units in a general definition.

(b) (i) Change $= 0.36 \times 9.6 = 3.46\,\text{kg}\,\text{m}\,\text{s}^{-1}$

(ii) Force $= 3.46\,\text{N}$

Mass to be lost $= 650 - (3.46 \times 9.8)$
$\qquad\qquad\qquad = 616\,\text{g}$

$\text{Time} = \dfrac{616}{360} = 1.7\,\text{s}$

 3 marks. Part (i) is correct. In part (ii) the candidate realises that the force is the rate of change of momentum (1 mark). However, when finding the mass to be lost, g has been used wrongly and the mass has been left in grams. The final calculation of the time from the candidate's figure for the mass loss is correct (1 mark).

(c) Because the mass is changing and the force is constant, the acceleration changes.

 1 mark only. This is an error frequently made by candidates. Although all factors have been mentioned, the direction of any change has not been included. Also, the formula $F = ma$ has been assumed (not stated).

Question 10.2

Two spheres A and B of mass 100 g and 150 g, respectively, are suspended from light flexible threads so that, when the spheres are hanging vertically, they just touch, as shown in part (i) of the diagram below.

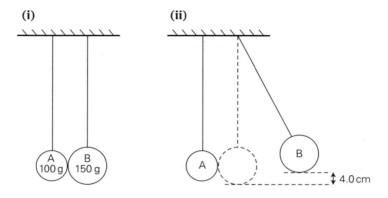

Sphere B is then moved to one side, keeping its thread taut, so that the sphere rises vertically by 4.0 cm, as shown in part (ii) of the diagram.

(a) Sphere B is released. Use energy considerations to determine its speed when it comes into contact with sphere A. The acceleration of free fall is 9.8 m s⁻².

(3 marks)

(b) When the spheres collide, they stick together. Determine:
(i) the speed of the spheres immediately after contact
(ii) the vertical height through which they rise

(5 marks)

(c) Suggest, with a reason, whether the collision in part **(b)** of the question is elastic or inelastic.

(2 marks)

(d) In another similar arrangement of spheres, the two spheres suspended from the threads are identical and perfectly elastic. One sphere is pulled to one side and then released. Describe the subsequent motion of both spheres.   (3 marks)

Total: 13 marks

Answer to Question 10.2: candidate A

(a) Loss in P.E. = gain in K.E.

$$mgh = \tfrac{1}{2}mv^2$$

$$v = \sqrt{2 \times 9.8 \times 0.04}$$

$$= 0.89 \, \text{m s}^{-1}$$

e 3 marks. Correct and explained well.

(b) (i) Momentum before = momentum after

$150 \times 0.89 = (150 + 100) \times v$

$v = 0.534 \, \text{m s}^{-1}$

(ii) Height $= \dfrac{v^2}{2g}$

$= 1.45 \, \text{cm}$

🅔 5 marks. It would have been better if the candidate had worked in kilograms in part (i). However, because mass appears on both sides of the equation and grams are used consistently, no harm has been done. In part (ii) explanation is lacking but it is assumed that the equation in part (a) (i) has been used.

(c) Inelastic because velocity of separation is not equal to velocity of approach.

🅔 2 marks. The suggestion is correct and the candidate shows a clear understanding of the condition for an elastic collision.

(d) B hits A, and B stops moving. Sphere A then moves off. When it returns and hits B, A stops and B moves off. The process continues.

🅔 2 marks. A mark has been lost because the candidate did not state that the one sphere would move off with the same speed as it was hit by the other.

Answer to Question 10.2: candidate B

(a) $mgh = \tfrac{1}{2}mv^2$

$v = \sqrt{2 \times 9.8 \times 0.04}$

$= 0.89 \, \text{m s}^{-1}$

🅔 3 marks. Correct but the explanation is barely adequate.

(b) (i) Momentum before = momentum after

$0.15 \times 0.89 = 0.15v + 0.10v$

$v = 0.534 \, \text{m s}^{-1}$

(ii) Height $= \dfrac{v^2}{2g}$

$= 14 \, \text{cm}$

🅔 3 marks. Two marks have been deducted in part (ii). To two significant figures, the answer should be 0.015 m. The candidate has made an error when converting metres to centimetres. Also, either the 'rounding down' of the answer is faulty or g has been taken as $10 \, \text{m s}^{-2}$. Always use the given value of any constant.

(c) Inelastic because they stick together.

🅔 1 mark. The suggestion is correct but the candidate has given no explanation other than the statement in the question. When giving an explanation, there are no marks for repeating what is given in the question.

(d) B hits A and B stops moving. Sphere A then moves off.

 1 mark. This is a very common mistake: only the first collision has been considered when the subsequent motion of both spheres was required. Also, the magnitudes of the speeds have not been mentioned.

Quantum physics

Question 11.1

(a) Explain what is meant by a *photon*. (3 marks)

(b) Light of wavelength 450 nm is incident on a metal surface. The minimum energy required to eject an electron from the surface of the metal is 3.5×10^{-19} J. At the surface of the metal, the light has an intensity of $0.46\,W\,m^{-2}$.
   (i) On the basis of the wave theory of light, calculate the time for an electron to obtain sufficient energy to be ejected. You may assume that the electron can collect all the light power in an area equal to the 'cross-sectional area' of the atom, that is $2.5 \times 10^{-18}\,m^2$. (3 marks)
   (ii) Comment on your answer to (i) with reference to experimental observations of the photoelectric effect. (2 marks)
   (iii) Determine the maximum kinetic energy with which an electron is emitted. The Planck constant $= 6.63 \times 10^{-34}$ J s; the speed of light $= 3.00 \times 10^8\,m\,s^{-1}$. (3 marks)
   (iv) Calculate the number of electrons ejected per second from an area of $1.5\,cm^2$ given that 5% of incident photons cause photoelectric emission. (3 marks)

Total: 14 marks

■ ■ ■

Answer to Question 11.1: candidate A

(a) A photon is a packet of energy of electromagnetic radiation. Its energy is given by *hf*.

> 🄴 2 marks. It would be better to refer to a quantum rather than a packet. The symbols *hf* should have been explained.

(b) (i) Energy per second gathered by electron = intensity × area
$$= 0.46 \times 2.5 \times 10^{-18}$$
$$= 1.15 \times 10^{-18}\,J$$

$$\text{time} = \frac{3.5 \times 10^{-19}}{1.15 \times 10^{-18}} = 0.30\,s$$

(ii) This time is quite short but, according to the photoelectric effect theory, the electron is ejected as soon as the photon hits it. So the wave theory cannot explain it.

(iii) For the photoelectric effect
photon energy = work function energy + kinetic energy

$$\text{photon energy} = \frac{hc}{\lambda} = \frac{6.63 \times 10^{-34} \times 3.00 \times 10^8}{450 \times 10^{-9}}$$

$$= 4.42 \times 10^{-19}\,J$$

$4.42 \times 10^{-19} = 3.5 \times 10^{-19} + \text{K.E.}$

$\text{K.E.} = 9.2 \times 10^{-20}\,\text{J}$

(iv) Number of photons per second $= \dfrac{0.46}{4.42 \times 10^{-19}}$

$$= 1.04 \times 10^{18}$$

Number of electrons per second $= 1.04 \times 10^{18} \times 0.05$

$$= 5.2 \times 10^{16}$$

9 marks. Full marks have been scored in part (i). In part (ii) it would have been more correct to write about the instantaneous emission of the photoelectron as being an observation, rather than an outcome of photoelectric theory. The calculation in part (iii) is explained well and is correct. In part (iv) the number ejected per second from an area of $1\,\text{m}^2$, not $1.5\,\text{cm}^2$, has been found. The correct answer is 7.8×10^{12}.

■ ■ ■

Answer to Question 11.1: candidate B

(a) It is a packet of light energy.

1 mark. The candidate should have mentioned electromagnetic radiation, rather than light. Also, a formula for the energy of a photon has not been given.

(b) (i) Energy per second gathered by electron = intensity $\times$ area

$$= 0.46 \times 2.5 \times 10^{-18}$$
$$= 1.15 \times 10^{-18}\,\text{J}$$

$\text{Time} = \dfrac{1.15 \times 10^{-18}}{3.5 \times 10^{-19}} = 3.28\,\text{s}$

(ii) The wave theory can't be right because this is a long time for the electron to be ejected.

(iii) For the photoelectric effect

$\text{frequency} = \dfrac{3.00 \times 10^8}{450\,\text{nm}} = 6.67 \times 10^{15}\,\text{Hz}$

photon energy $= hf = 6.63 \times 10^{-34} \times 6.67 \times 10^{15}$
$$= 4.42 \times 10^{-18}\,\text{J}$$

$4.42 \times 10^{-18} = 3.5 \times 10^{-19} + \text{K.E.}$

$\text{K.E.} = 4.07 \times 10^{-18}\,\text{J}$

(iv) Number of photons per second $= \dfrac{0.46}{4.42 \times 10^{-18}}$

$$= 1.04 \times 10^{17}$$

Number of electrons per second on area $= 1.04 \times 10^{17} \times 0.05 \times 1.5$
$$= 7.8 \times 10^{15}$$

5 marks. In part (i) the calculation of the energy is correct, but the inverse of the time has been found. No marks have been scored in part (ii). A 'long time' has no

meaning here — a long time compared with what? The candidate should have mentioned that the electron is ejected instantaneously. In part (iii) the candidate appears not to know that 1 nm is equal to 10^{-9} m. The remainder of the calculation in part (iii) is correct, when based on the candidate's value for the photon energy. The area substituted into the equation in part (iv) should have been 1.5×10^{-4} m^2.

Question 11.2

(a) Outline an experiment by which the wave nature of particles may be demonstrated.

(6 marks)

(b) Light of wavelength 4.0×10^{-7} m and intensity 0.75 W m^{-2} is incident normally on a surface that absorbs the photons. The speed of light c is 3.0×10^{8} m s^{-1} and the Planck constant h is 6.6×10^{-34} J s. Calculate:
(i) the energy of a photon of the light
(ii) the number of photons incident per second on 1.0 m^2 of the surface
(iii) the momentum of a photon
(iv) the change in momentum per second per metre squared of the surface

(8 marks)

(c) Suggest why your answer to (b) (iv) is sometimes referred to as *radiation pressure*.

(3 marks)

Total: 17 marks

Answer to Question 11.2: candidate A

(a)

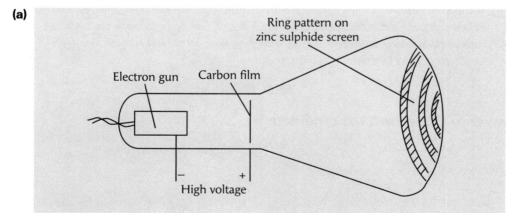

The electrons shoot out of the electron gun and pass through the carbon film. A number of rings are seen on the screen. This is a diffraction pattern, showing that particles can behave as waves. If the high voltage is increased, the rings have a smaller radius. This shows that the faster the electrons travel, the shorter the wavelength.

  🅮 5 marks. The candidate drew a diagram and indicated clearly the main features of the apparatus. A good diagram will usually save on the amount of writing required. The candidate has described the observations made and the conclusions reached. The candidate has made a common omission — there is no indication that there must be a vacuum in the apparatus.

(b) (i) Energy $= hf = \dfrac{hc}{\lambda} = \dfrac{6.6 \times 10^{-34} \times 3.0 \times 10^8}{4.0 \times 10^{-7}}$

$\qquad\qquad = 4.95 \times 10^{-19}$ J

(ii) Number $= \dfrac{\text{power}}{\text{energy}} = \dfrac{0.75}{4.95 \times 10^{-19}}$

$\qquad\qquad = 1.52 \times 10^{18}$ per second

(iii) $\lambda = \dfrac{h}{p}$

$\qquad$ momentum $= \dfrac{6.6 \times 10^{-34}}{4.0 \times 10^{-7}}$

$\qquad\qquad = 1.65 \times 10^{-27}$

(iv) Change $= 1.65 \times 10^{-27} \times 1.52 \times 10^{18}$

$\qquad\qquad = 2.5 \times 10^{-9}\,\text{s}^{-1}\,\text{m}^{-2}$

  🅮 6 marks. The calculations are correct with the formulae indicated. However, the unit in part (iii) should be either $\text{kg}\,\text{m}\,\text{s}^{-1}$ or $\text{N}\,\text{s}$ and in part (iv) it should be $\text{kg}\,\text{m}^{-1}\,\text{s}^{-2}$ or $\text{N}\,\text{m}^{-2}\,\text{s}^{-1}$.

(c) When momentum changes, a force is involved and this gives rise to pressure.

  🅮 1 mark. The answer is in very general terms. The candidate should have realised that insufficient detail has been given for 3 marks. Statements were required to indicate that rate of change of momentum is equal to force. This force is acting on unit area, and force per unit area is pressure.

■ ■ ■

Answer to Question 11.2: candidate B

(a)

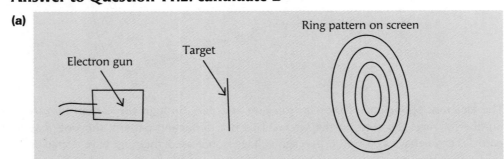

When the electron gun is switched on, a ring pattern is seen on the screen. This is a diffraction pattern showing that electrons can be diffracted.

e 2 marks. The diagram lacks detail. The glass envelope, the vacuum, the fluorescent screen and the accelerating voltage are not shown. Also, some detail of the target is required. An indication of the shape of the diffraction pattern is given, but the effect of change of electron speed has not been included.

(b) (i) Energy $= hf = \dfrac{hc}{\lambda} = \dfrac{6.6 \times 10^{-34} \times 3.0 \times 10^8}{4.0 \times 10^{-7}}$

$= 4.95 \times 10^{-19}\,\text{J}$

(ii) Number $= \dfrac{\text{power}}{\text{energy}} = \dfrac{0.75}{4.95 \times 10^{-19}}$

$= 1.52 \times 10^{18}$ per second

(iii) Momentum $= 6.6 \times 10^{-34} \times 4.0 \times 10^{-7}$

$= 2.64 \times 10^{-40}\,\text{N}$

(iv) Change $= 2.64 \times 10^{-40} \times 1.52 \times 10^{18}$

$= 4.0 \times 10^{-22}\,\text{N s}^{-1}$

e 4 marks. The calculations in parts (i) and (ii) are correct with the formulae indicated. In part (iii) no formula was given and it appears as if the candidate is using the formula $p = h\lambda$. This is, of course, incorrect. It may be that the candidate knew the formula $\lambda = \dfrac{h}{p}$ but made an error in the algebra before writing down any values. Always quote the formula used. It can make the difference between no marks (for 'bad physics') and some credit, having allowed for an error in the mathematics. The units in parts (iii) and (iv) are incorrect.

(c) Pressure is a force and the photons push on the area.

e No marks. Pressure is not a force. The candidate has not attempted to describe what is, in essence, the theory of how the atoms of an ideal gas exert a pressure on a surface.

Force fields

Question 12.1

(a) (i) Describe what is meant by an electric field.

(ii) Hence define electric field strength. (4 marks)

(b) A small metal sphere C is positively charged. It is placed near to an isolated metal sphere S, initially uncharged, as shown in the diagram below.

Copy the diagram, and on your copy show the distribution of charge on the sphere S. (2 marks)

(c) State and explain why there is no resultant electric field inside sphere S.

(4 marks)

Total: 10 marks

Answer to Question 12.1: candidate A

(a) (i) Region where a charge experiences a force.

(ii) Field strength = $\dfrac{\text{force}}{\text{charge}}$

🅔 2 marks only. In part (i) the fact that the charge must be stationary has not been included. If the charge were to be moving, the field could be magnetic. In part (ii) positive charge has not been mentioned.

(b)

🅔 2 marks. The induced charges are in approximately the correct positions and there are equal numbers of each type of charge.

(c) There cannot be a field because the field would make charges move. The induced charges cause a field to the left. This cancels out the field due to sphere C.

🅔 3 marks. It would be better to refer to the two fields as being equal and opposite in direction in the sphere S and to say that charges will move until there is no resultant field.

Answer to Question 12.1: candidate B

(a) **(i)** Area where a stationary charge experiences a force.
(ii) Field strength is the force on a unit positive charge.

> 🅴 2 marks only. It would be better to refer to a region of space, rather than an area. In part (ii), although positive charge has been mentioned, no marks have been scored because no ratio has been given and it appears as if field strength is a force.

(b)

> 🅴 1 mark only. The induced positive and negative charges are not equal in number.

(c) The charges in the sphere S would move if there was a field in the conductor. The induced charges produce a field which counteracts the field of the sphere C.

> 🅴 2 marks. Although it is clear that there is no field in S, the relationship between the field due to the charge on sphere C and that due to the induced charges, giving rise to no resultant field, is not clear.

Question 12.2

(a) State Newton's law of gravitation. (2 marks)

(b) Show that the field strength g at a distance r from a point mass M is given by the expression

$$g = \frac{GM}{r^2}$$

(3 marks)

(c) The field strength outside a uniform sphere of mass m is the same as that due to a point mass m at its centre. Make suggestions (one in each case) as to why:
(i) the field strength on the surface of the Moon is only one-sixth that on the Earth's surface although their mean densities are very similar
(ii) the value of g varies over the surface of the Earth
(iii) the value of g at the surface of the Earth and at 1000 m above its surface are almost equal

(8 marks)

Total: 13 marks

■ ■ ■

Answer to Question 12.2: candidate A

(a) For two point masses, the force between them is proportional to the product of their masses and inversely proportional to their separation.

🅔 2 marks — a good answer. Reference has been made to the fact that the law applies to point masses.

(b) Force $= \dfrac{GMm}{r^2}$ and $F = mg$

Therefore, $mg = \dfrac{GMm}{r^2}$

$g = \dfrac{GM}{r^2}$

🅔 2 marks. Although the algebra is correct, the working lacks explanation. It should be stated that m is a small mass, distance r from the mass M.

(c) (i) Since mass $= \frac{4}{3}\pi r^3 \rho$
$g = G \times \frac{4}{3}\pi r^3 \rho \div r^2$
$\quad = G \times \frac{4}{3} \times \pi r \rho$
The radius of the Moon is less than the Earth's and so g will be less on the Moon.
(ii) The radius of the Earth is not the same everywhere and so g will vary.
(iii) If the radius of the Earth is r, then the new distance is $(r + 1000)$ and the new field strength is

$$\frac{GM}{(r + 1000)^2}$$

But r is much greater than 1000 m and so r is almost the same as (r + 1000). Hence the values of g are almost equal.

🅔 7 marks. The answer in part (i) is very comprehensive and scores three marks. In part (ii) the candidate understands that the distance from the surface to the centre is not constant but has not actually related this fact to the equation for g. The spin of the Earth has not been considered. The fact that 1000 m is small compared with the radius of the Earth is clear and the consequence of this on the value of g has been discussed by reference to the formula for g.

■ ■ ■

Answer to Question 12.2: candidate B

(a) For two masses M and m separated by distance r, $F = \dfrac{GMm}{r^2}$.

🅔 1 mark only. It has not been stated that the law applies to point masses. Also, it would be preferable to explain the symbols F and G.

(b) $mg = \dfrac{GMm}{r^2}$

$g = \dfrac{GM}{r^2}$

🅔 1 mark. This is a typical situation in which it appears that the candidate has learned the work but gives no explanation, thus losing marks. Questions asking candidates to 'show that' will inevitably carry marks for explanation.

(c) (i) The mass of the Moon is less and so g will be smaller.

(ii) Some of the gravitational force provides the centripetal force because the Earth is spinning. Near the equator, the centripetal force is greater than at the poles. The value of g will be greater at the poles.

(iii) The radius of the Earth is much greater than 1000 m and so GM/r^2 will not change.

3 marks. No marks have been scored in part (i) — the mass may be smaller, but so is the radius, and g depends on r^{-2}! The candidate should have realised that density would be involved because density is mentioned in the question. In part (ii) the basic idea is given, but the answer lacks clarity. The relation mg + centripetal force = gravitational force should have been included. The fact that 1000 m is small compared with the radius of the Earth is made clear in part (iii) and a reference to the formula for g has been made. However, some comment should have been made about the effect on the formula. For example, 'GM/r^2 will be almost the same as $GM/(r + 1000)^2$'.

Question 12.3

(a) A planet has a radius r and the acceleration of free fall at its surface is g. A satellite, mass m, is placed in a circular orbit of radius R round the planet. The period of rotation of the satellite round the planet is T.

(i) Write down an expression for the linear speed v of the satellite in terms of T and R.

(ii) Show that T and R are related by the expression

$$T^2 = \frac{(4\pi^2 R^3)}{gr^2}$$
(5 marks)

(b) Data for some of the moons of the planet are given in the table below.

Moon	Radius/km	Period/days
A	1.81×10^5	0.50
B	6.71×10^5	3.55
C	1.88×10^6	16.7
D	2.12×10^7	631

(i) Use the data to verify that $T^2 \propto R^3$.

(ii) A further moon is discovered orbiting the planet with a period of 0.67 days. Determine the radius of its orbit.
(8 marks)

Total: 13 marks

Answer to Question 12.3: candidate A

(a) (i) $v = \dfrac{2\pi R}{T}$

(ii) At the surface of the planet, $g = \dfrac{GM}{r^2}$

For the moon, $\dfrac{mv^2}{R} = \dfrac{GMm}{R^2}$

$\dfrac{v^2}{R} = \dfrac{gr^2}{R^2}$

$\therefore v^2 = \dfrac{gr^2}{R}$

but $v^2 = \dfrac{4\pi^2 R^2}{T^2}$

$\dfrac{4\pi^2 R^2}{T^2} = \dfrac{gr^2}{R}$

$4\pi^2 R^3 = gr^2 T^2$

therefore, $T^2 \propto R^3$ because g and r are constant.

e 5 marks. Part (i) requires no working. Part (ii) is a well-recognised derivation. The candidate has set out the work clearly. Note that the use of the expression $gr^2 = GM$ as a means of eliminating GM from the formula is a common practice.

(b) (i)

Moon	$\dfrac{R^3}{T^2}$
A	2.37×10^{16}
B	2.39×10^{16}
C	2.38×10^{16}
D	2.39×10^{16}

To three significant figures, $\dfrac{R^3}{T^2}$ is constant. That is, $T^2 \propto R^3$.

(ii) Average value of $\dfrac{R^3}{T^2} = 2.38 \times 10^{16}$

For moon with $T = 0.67$ days
$R^3 = 2.38 \times 10^{16} \times 0.67^2$
$R = 2.2 \times 10^5$ km

e 8 marks. Calculations similar to this require care with a calculator, but it should not be difficult to score high marks. The candidate did not give a unit for $\dfrac{R^3}{T^2}$ but this was not necessary (although perhaps advisable).

Answer to Question 12.3: candidate B

(a) (i) $v = \dfrac{2\pi R}{T}$

(ii) Centripetal force = gravitational force
$\dfrac{mv^2}{R} = \dfrac{GMm}{R^2}$

$$\frac{4\pi^2 R^2}{T^2 R} = \frac{GM}{R^2}$$

$$4\pi^2 R^3 = GMT^2$$

e 3 marks. Part (i) is correct. In part (ii) the candidate has not used the expression $gr^2 = GM$ to eliminate GM from the formula. It is not correct to state that the centripetal force is equal to the gravitational force. In fact, the gravitational force *provides* the centripetal force.

(b) (i)

Moon	LogR	LogT
A	5.258	−0.301
B	5.827	0.550
C	6.274	1.223
D	7.326	2.800

Plot a graph of log$R$ against log$T$.

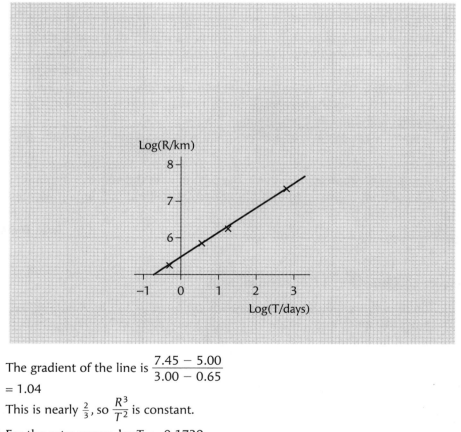

The gradient of the line is $\dfrac{7.45 - 5.00}{3.00 - 0.65}$

$= 1.04$

This is nearly $\frac{2}{3}$, so $\dfrac{R^3}{T^2}$ is constant.

(ii) For the extra moon, log$T = -0.1739$

From the graph, log$R = 5.35$

$R = 2.2 \times 10^5$ km

e 3 marks. The candidate plotted a correct graph but the scale is too small. For this graph, the gradient is about 0.67. There is an error when reading the co-ordinates of one of the points for the calculation of the gradient. Part (ii) is correct but the small scale of the graph could have meant that the result lacks reliability.

Question 12.4

(a) Two small positively charged spheres are a large distance apart. They are then brought closer together so that they nearly touch.
  (i) Explain why work has to be done to bring the spheres closer together.
  (ii) Suggest why this work done may be referred to as electric potential energy.   (4 marks)

(b) The electric potential energy E_P of two point charges Q and q separated by a distance r is given by the expression

$$E_P = 8.9 \times 10^9 \times \frac{Qq}{r}$$

A proton of mass 1.67×10^{-27} kg and radius 1.2×10^{-15} m is brought towards a lithium nucleus $^{7}_{3}$Li of radius 2.3×10^{-15} m until they just touch. The elementary charge is 1.6×10^{-19} C and the charges on the proton and the lithium nucleus may be assumed to be point charges.
  (i) Calculate the electric potential energy of the proton when it is just in contact with the lithium nucleus.
  (ii) Determine the initial speed of the proton so that it just reaches the lithium nucleus.   (5 marks)

(c) The equipment required to provide the proton with the energy calculated in **(b)** is sometimes referred to as a high-energy accelerator. Suggest why it is called *high-energy*.   (2 marks)

Total: 11 marks

■ ■ ■

Answer to Question 12.4: candidate A

(a) **(i)** The spheres repel each other and work is done against this force.
     (ii) When the spheres are released, they will move apart and could be made to do work. That is, they have the potential to do work.

e 3 marks. It is unfortunate that, in part (i), the candidate did not refer to *electrostatic* repulsion. Also, in part (ii), although potential energy has been well explained, the reason for it being *electric* potential energy has not been discussed.

(b) **(i)** Energy $= 8.9 \times 10^9 \times \dfrac{(1.6 \times 10^{-19})^2 \times 3}{3.5 \times 10^{-15}}$

$= 1.95 \times 10^{-13}$ J

(ii) $\frac{1}{2}mv^2 = 1.95 \times 10^{-13}$

$\frac{1}{2} \times 1.67 \times 10^{-27} \times v^2 = 1.95 \times 10^{-13}$

$v = 1.53 \times 10^7 \, \text{m s}^{-1}$

🅔 5 marks. In part (i) a common mistake of failing to square the term (1.6×10^{-19}) has been avoided. Note that the substitutions into formulae are clear.

(c) The energy is more than 1 MeV and so, on a nuclear scale, this energy is large.

🅔 2 marks. The candidate has made the important point that the energy is large on a nuclear scale.

■ ■ ■

Answer to Question 12.4: candidate B

(a) (i) Like charges repel and so you do work to bring them together.
(ii) Potential energy is energy due to position and the spheres are close together.

🅔 2 marks. In part (i) the candidate has scored both marks. However, in part (ii) no marks have been given. There is no reference to the ability to do work as the spheres move apart due to the electrostatic force of repulsion.

(b) (i) Energy $= 8.9 \times 10^9 \times \dfrac{(1.6 \times 10^{-19})^2}{1.2 \times 10^{-15}}$

$= 1.9 \times 10^{-13} \, \text{J}$

(ii) $\frac{1}{2} \times 1.67 \times 10^{-27} \times v^2 = 1.95 \times 10^{-13}$

$v = 1.51 \times 10^7 \, \text{m s}^{-1}$

🅔 2 marks. In part (i), although the answer appears to be correct, there are two errors. First, the charge on the lithium nucleus has been taken as e, and second, the final separation is incorrect. In part (ii) the substitution and answer are correct, allowing for the candidate's answer to part (i).

(c) They are called high-energy accelerators because a lot of energy has to be put into them.

🅔 No marks. The candidate does not appear to have any idea of the situation. Always read the question carefully. The examiners had given a clear indication as to how to proceed. That is, refer to the energy calculated in (b).

Magnetic effects

Question 13.1

(a) (i) Draw a diagram to show the magnetic field near to a long straight current-carrying wire.

(ii) State and explain a rule that may be used to determine the direction of the magnetic field in (i). (5 marks)

(b) Two long straight wires A and B are placed parallel to one another, as illustrated below.

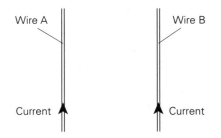

A current is passed through both wires in the upward direction.

(i) Explain why there is a force acting on both wires and predict its direction.

(ii) Explain why this effect may be used as a basis of the definition of the amp. (5 marks)

Total: 10 marks

Answer to Question 13.1: candidate A

(a) (i)

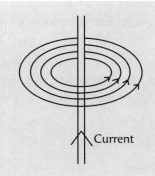

(ii) The right-hand grip rule. The wire is grasped in the right hand with the thumb pointing in the direction of the current. The fingers give the direction of the magnetic field.

e 4 marks. The diagram shows concentric circles in a plane normal to the wire. The direction is correct but the spacing of the lines should increase with increasing distance from the wire. The rule is described well.

(b) (i)

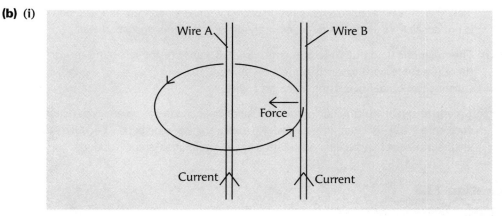

The magnetic field due to current in wire A is at right angles to wire B. By Fleming's left-hand rule there will be a force on wire B towards wire A. Newton's third law states that there will be an equal but opposite reaction on wire A. So the wires try to move together.

(ii) The force on the wires depends on their length, distance apart and the current when the wires are in a vacuum. Therefore current can be defined from the fundamental quantities because force and lengths can be related to the fundamental quantities.

e 5 marks. A good answer that includes all the points. Note that the drawing of a sketch diagram helps enormously with the explanation.

■ ■ ■

Answer to Question 13.1: candidate B

(a) (i)

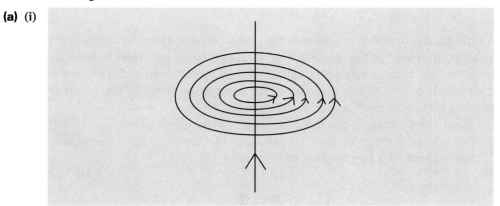

(ii) Imagine a right-hand corkscrew screwed in the direction of the current. The motion of the thumb gives the field direction.

🅔 2 marks. Concentric circles are evident but in which plane? The diagram appears 'flat'. The lines do not show increasing spacing with distance from the wire and it is not possible to judge the direction of the field. The rule is described well.

(b) (i) The magnetic field due to wire A cuts wire B. Because of the left-hand motor rule, there is a force on B towards A.

(ii) Current can be defined using force on a wire.

🅔 1 mark. In part (i) the direction of the field is not clear and the force on wire A is not considered. A sketch may have helped. No marks are scored in part (ii) because the idea that current may be defined in terms of base quantities is not included.

Question 13.2

(a) Explain what is meant by:
(i) a magnetic field
(ii) magnetic flux density   (4 marks)

(b) A horseshoe magnet, as illustrated, is placed on the pan of a balance.

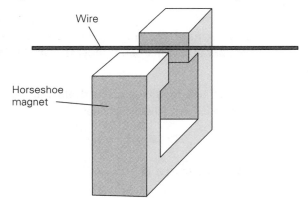

Wire

Horseshoe magnet

A rigid wire is held firmly between the poles of the magnet so that it is at right angles to the field of the magnet. The magnet may be assumed to produce a uniform field between its poles, and zero field everywhere else. The length of the poles is 6.0 cm. When a current of 5.0 A is passed through the wire, the reading on the balance is seen to *increase* by 0.49 g.
(i) State, with a reason, the direction of the force on the wire.
(ii) Calculate the magnitude of the magnetic flux density between the poles of the magnet. The acceleration of free fall is 9.8 m s^{-2}.   (6 marks)
                                                                    Total: 10 marks

Answer to Question 13.2: candidate A

(a) (i) A region where a current-carrying wire will experience a force.

(ii) The strength of a magnetic field. A wire carrying a current I in a magnetic field experiences a force F given by $F = BIL$ when the current is at right angles to the field. Flux density is defined as $\frac{F}{IL}$ where L is the length of the wire.

 4 marks. The candidate has identified each term and has explained how flux density may be defined. Note that the question does not ask for units.

(b) (i) Force on magnet is downwards so force on wire is upwards.

(ii) $F = BIL$

$0.49 \times 10^{-3} \times 9.8 = B \times 5.0 \times 6.0 \times 10^{-2}$

$B = 0.016\,\text{T}$

 5 marks. In part (i) the candidate did not mention Newton's third law or that action and reaction are equal but opposite. Part (ii) is correct, with correct conversion of balance reading to newtons and length to metres. The unit could have been given as $\text{Wb}\,\text{m}^{-2}$.

■ ■ ■

Answer to Question 13.2: candidate B

(a) (i) A region where a magnet feels a force.

(ii) The strength of a magnetic field, measured in teslas.

 Only 1 mark. The explanation in part (i) is not appropriate at AS/A-level. It should be based on the means by which the magnitude of the field strength is determined. In part (ii) a statement is given, rather than an explanation, as expected by the question.

(b) (i) Force is upwards.

(ii) $F = BIL$

$0.49 \times 9.8 = B \times 5.0 \times 6.0 \times 10^{-2}$

$B = 16\,\text{T}$

 2 marks. In part (i) the answer may well be guesswork. Where examiners expect a reason for the answer and the answer (by itself) has only two options, then no mark is awarded unless some reasoning is given. Be warned! In part (ii) the balance reading has not been converted correctly to newtons, but otherwise the calculation is correct and would be awarded two of the three available marks. A mark would have been deducted for no unit or a wrong unit.

Question 13.3

(a) (i) State Lenz's law of electromagnetic induction.

(ii) Explain why it is a consequence of the law of conservation of energy.

(4 marks)

(b) Two coils A and B are constructed using insulated wire. Coil B is wound round the centre of coil A, as illustrated in the diagram below.

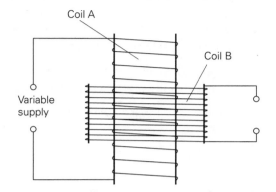

Coil A is connected to a variable supply and the current *I* in coil A varies with time *t* as shown in the graph below.

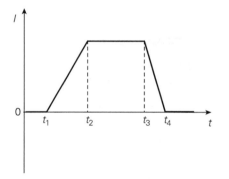

An emf *E* is induced in coil B. Draw a graph to show the variation with time *t* of the emf *E*. Make sure that you write the relevant times on your graph.

(4 marks)

Total: 8 marks

Answer to Question 13.3: candidate A

(a) (i) Lenz's law states that the induced emf acts in such a direction as to produce effects to oppose the change causing it.

(ii) The induced current is energy and that energy must have come from somewhere. Work is done opposing the change and this work becomes electrical energy.

> e 4 marks. An accurate statement in part (i), and in part (ii) the energy transformation has been made clear.

(b)

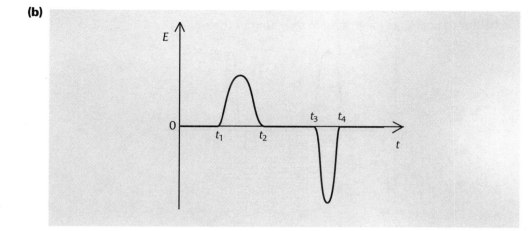

3 marks. The shape of the peaks has been penalised. The peaks should have opposite polarities, but whether the first should be positive or negative cannot be deduced unless the directions of the windings on the coils are made clear. The correct solution is shown below.

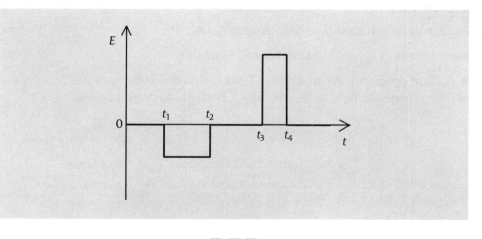

■ ■ ■

Answer to Question 13.3: candidate B

(a) (i) The induced emf opposes the motion producing it.

(ii) Work has to be done to produce electrical energy, such as in a dynamo.

1 mark. No marks are scored in part (i). The emf produces effects to oppose the change. An emf cannot, for example, oppose the motion of a magnet. Furthermore, motion need not occur, but there is always a change in flux linkage. In part (ii), 1 mark has been scored. No indication has been given as to how the work is done.

(b)

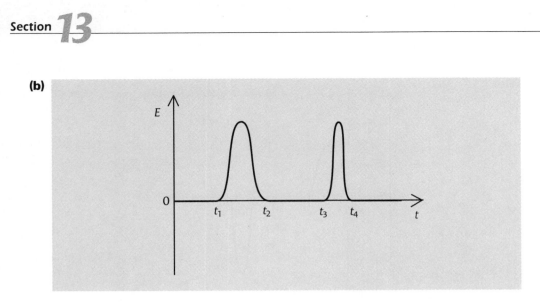

🅔 1 mark only for the fact that there is no induced emf when the current is constant. The peaks are the wrong shape, they shouldn't have the same polarity and they shouldn't be the same height.

Question 13.4

(a) Explain what is meant by a change in magnetic flux.   (2 marks)

(b) State Faraday's law of electromagnetic induction.   (2 marks)

(c) A straight conductor XY of length 8.0 cm is situated at right angles to a uniform magnetic field of flux density 3.2 mT, as illustrated below.

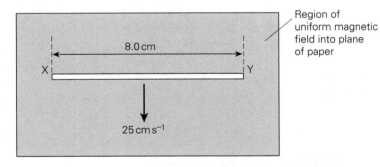

The wire is moved at right angles to the magnetic field at a constant speed of 25 cm s⁻¹.
(i) State and explain which end of the wire, X or Y, is positive with respect to the other.
(ii) Calculate the emf induced between the ends of the wire.   (6 marks)

Total: 10 marks

■ ■ ■

Answer to Question 13.4: candidate A

(a) When either magnetic field changes or when a conductor moves in a constant magnetic field.

> 🄴 2 marks. It would have been better to refer to magnetic field strength, rather than just magnetic field.

(b) The induced emf is proportional to the rate of change of flux linkage.

> 🄴 2 marks. The candidate has referred to emf (not current) and to rate of change of flux linkage.

(c) (i) Using Fleming's right-hand rule, current flows towards Y. This means Y is negative.

(ii) $E = Blv$

$$= 3.2 \times 10^{-3} \times 8.0 \times 10^{-2} \times 25 \times 10^{-2}$$

$$= 64\,\mu V$$

> 🄴 5 marks. The right-hand rule has been used correctly, but it would have been better to talk about conventional current in this case. However, thinking about the wire as a source of emf, then Y should be *positive*. The calculation is correct.

■ ■ ■

Answer to Question 13.4: candidate B

(a) Whenever the size of the magnetic field changes.

> 🄴 1 mark. There is no mention of movement within a field of constant strength. It is better to refer to strength of a field, rather than size.

(b) Induced current $= \dfrac{-\mathrm{d}(N\phi)}{\mathrm{d}t}$

> 🄴 No marks. The candidate should refer to induced emf, not induced current. There will be an induced emf when the flux linkage changes but there will be an induced current only if there is a complete circuit. The expression $\dfrac{-\mathrm{d}(N\phi)}{\mathrm{d}t}$ may be correct, but there is no explanation of the symbols used.

(c) (i) The induced current flows from X to Y so X is positive.

(ii) Area swept out $= 8 \times 25 = 200\,cm^2$

emf $= B \times$ area

$$= 3.2 \times 10^{-3} \times 200 \times 10^{-2}$$

$$= 6.4 \times 10^{-3}\,V$$

> 🄴 3 marks. The rule has not been named and the polarity is incorrect. One mark has been awarded for the direction of the induced current. In part (ii) the calculation is correct apart from the conversion of cm^2 to m^2. The fact that the candidate has not referred to area swept out *per second* has been condoned, rather generously.

Capacitance

Question 14.1

(a) Outline an experiment to show the variation with time of the current in a resistor as a capacitor discharges through the resistor. (5 marks)

(b) A capacitor of capacitance C is discharged through a resistor of resistance R. The variation with time t of the potential difference V across the capacitor is shown in the graph below.

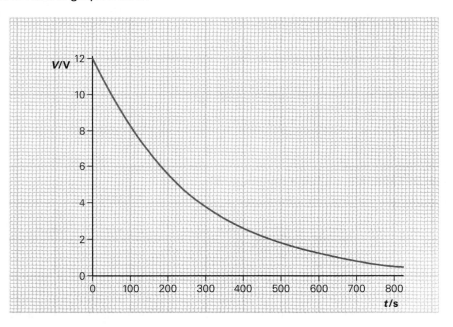

(i) Use the graph to determine the time constant of the circuit.
(ii) The resistor has resistance $420 \, k\Omega$. Calculate the capacitance C, in microfarads, of the capacitor.
(iii) Estimate the charge that flows off one plate of the capacitor during the first 100 seconds of its discharge. Explain your working. (8 marks)

Total: 13 marks

Answer to Question 14.1: candidate A

(a) Set up the circuit as shown below. Put the switch into position A. Move the switch to position B, start the clock and record the current. Take readings of the current every 10 seconds. Plot a graph of current on the y-axis against time.

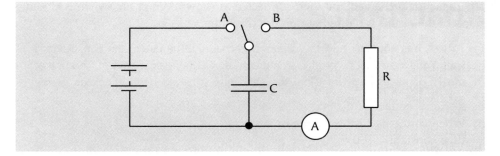

4 marks. There is adequate explanation of the procedure and how the results are presented. One mark has been lost because an ammeter is not appropriate for current measurement. A milliammeter or microammeter is required.

(b) (i) Time constant when voltage is $12 \times e^{-1} = 4.42\,\text{V} = 260\,\text{s}$
two time constants when voltage is $12 \times e^{-2} = 1.62\,\text{V} = 520\,\text{s}$
average time constant $= 260\,\text{s}$

(ii) Time constant $= CR$
$C \times 420 \times 10^3 = 260$
$C = 619\,\mu\text{F}$

(iii) Average voltage during first 100 seconds $= 10.1\,\text{V}$
average current $= \dfrac{V}{R} = 24\,\mu\text{A}$
charge $= It = 2.4 \times 10^{-5} \times 100 = 2.4 \times 10^{-3}\,\text{C}$

7 marks. The calculations are correct. The explanation in part (i) is clumsy (a voltage cannot equal a time) but would be condoned. Note that values for the time constant have been found twice and an average calculated. This is good practice at AS/A-level, and should also be used for half-life determination. However, further explanation was expected for part (iii). The candidate should have stated that it was assumed that the graph line is straight during the first 100 seconds.

Answer to Question 14.1: candidate B

(a)

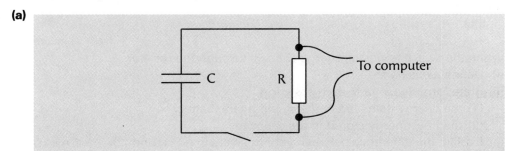

The capacitor is charged and then connected as shown. When the switch is closed, a graph of the results is plotted out on the computer.

e 1 mark. It is not clear how the capacitor is charged. The use of the IT package is vague and may well be producing a graph of voltage against current. The use of the IT package does not allow the candidate to explain the principles of the experiment. Be warned!

(b) (i) When voltage is $12 \times e^{-1}$, time is 260 s
time constant = 260 s
(ii) $CR = 260$
$C \times 420 \times 10^3 = 260$
$C = 6.19 \times 10^{-4}\,F$
(iii) Number of 1 cm squares = 20
charge = $20 \times 50 \times 1 = 1000\,C$

e 3 marks. In part (i) there has been no attempt to find a second value and then to average. Part (ii) is correct but the candidate has not answered the question by giving the answer in microfarads. Answer the question! In part (iii) the candidate realises that an area under a graph is involved, but the graph should be variation of the current with time. The candidate has failed to convert values of voltage to current by dividing by the resistance.

Question 14.2

(a) A capacitor of capacitance C is connected is series with a battery of emf E, a switch and a resistor of resistance R, as shown below.

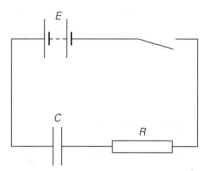

When the switch is closed, charge Q flows through the resistor.
(i) Define *capacitance*.
(ii) State formulae, one in each case, for:
    (1) the energy delivered to the circuit by the battery
    (2) the final energy stored in the capacitor
(iii) Explain the difference between your answers in (ii) (1) and (2).   (6 marks)

(b) A number of capacitors are marked '50 μF, 25 V'.

   (i) Explain what is meant by this marking.

   (ii) Draw diagrams, one in each case, to show how you would connect a number of these capacitors to produce arrangements having the following specifications:

     (1) 100 μF, 25 V

     (2) 75 μF, 50 V

                                                                             (4 marks)

Total: 10 marks

Answer to Question 14.2: candidate A

(a) (i) It is the ratio of the charge on an object to its potential.

   (ii) (1) Energy $= QE$

      (2) Energy $= \frac{1}{2}QE$

   (iii) Some energy has been lost in the resistor.

> 🄴 5 marks. In part (i) the capacitance of an isolated conductor has been defined. This is acceptable because no particular arrangement was specified. Part (ii) is correct. The answer in part (iii) is too trivial for AS/A-level. It was expected that the candidate would state that some electrical energy had been converted into heat in the resistor.

(b) (i) It means that the capacitance is 50 μF and that the capacitor can be used with a pd across it of up to 25 V.

   (ii)

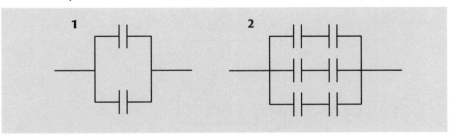

> 🄴 4 marks. The explanation in part (i) is satisfactory and the circuit diagrams in part (ii) are correct.

Answer to Question 14.2: candidate B

(a) (i) Capacitance is the charge required to cause unit increase in potential.

   (ii) (1) Energy $= QE$

      (2) Energy $= \frac{1}{2}QV$

   (iii) Heat is produced in the resistor.

🄔 2 marks. No mark is scored in part (i), because the ratio is not clear and it would appear as if capacitance is charge! Part (ii) (1) is correct, but in part (ii) (2) the symbols used in the formula are not all the same as those given in the question. This is a common mistake. The statement in part (iii) is correct but incomplete.

(b) (i) The $50\,\mu F$ capacitor can be used on a $25\,V$ supply.

(ii)

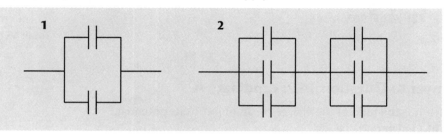

🄔 3 marks. The candidate has not made it clear that the *maximum* voltage that can be applied across the capacitor is $25\,V$. Part (ii) is correct. There are two $150\,\mu F$ arrangements, each with a potential difference of $25\,V$, giving the pd across the whole arrangement as $50\,V$.

Question 14.3

(a) A capacitor of capacitance C is charged to a potential difference V_0. It is then connected in series with a resistor of resistance R and a switch, as shown in the diagram below.

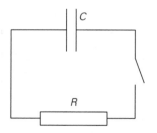

(i) Explain why, although the resistance of the capacitor is infinite, there is a current in the resistor when the switch is closed.

(ii) Sketch a graph to show the variation of the pd across the capacitor with time t after the switch has been closed.
(6 marks)

(b) The capacitor in **(a)** has capacitance $200\,\mu F$ and the resistance of the resistor is $6.0\,M\Omega$. Calculate the time taken for the capacitor to discharge 90% of its energy.
(4 marks)

Total: 10 marks

Answer to Question 14.3: candidate A

(a) (i) Charge moves from one plate to the other through the resistor. Current is charge in motion and charge flows through the resistor, not the capacitor.

(ii)

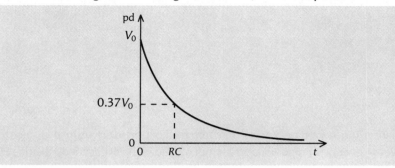

> 6 marks. A good answer in part (i). The sketch in part (ii) has the correct shape, starting at V_0. $V = 0.37V_0$ when $t = RC$ has also been shown. Note that this is a sketch: accurate plotting is not necessary.

(b) Energy $= \frac{1}{2}CV^2$
so since $V = V_0 e^{-t/RC}$

$$\frac{E_2}{E_1} = (e^{-t/RC})^2 \text{ and } RC = 1200$$

$$\frac{90}{100} = e^{-2t/1200}$$

$$t = 63\,s$$

> 3 marks. The calculation has been explained adequately but one mark has been lost because the time given in the answer is the time for the capacitor to discharge 10% of its energy, not the time taken to discharge 90% of its energy with 10% remaining. The correct answer is 1380 s.

■ ■ ■

Answer to Question 14.3: candidate B

(a) (i) Charge moves from one plate to the other and moving charge is a current.

(ii)

🄴 2 marks. In part (i) the candidate has the idea that moving charge is a current, but has not made it clear in which part of the circuit the charge is moving. The graph shows the general form of the relationship without any relevant features being marked. Remember when sketching a graph that important features should be indicated.

(b) $V = V_0 e^{-t/RC}$

$$\frac{90}{100} = e^{-t(200\ \mu F \times 6\ M\Omega)}$$

$$0.9 = e^{-t/1200}$$

$$t = 126\,s$$

🄴 2 marks. The candidate knows the formula for the discharge of a capacitor. However, energy is proportional to V^2, not V. Also, if the capacitor discharges 90% of its energy, 10% is remaining. One mark was given for the correct manipulation of the formula.

Nuclear physics

Question 15.1

(a) Explain what is meant by alpha decay (α-decay).   (3 marks)

(b) A stationary radium nucleus $^{226}_{88}$Ra undergoes radioactive decay to become a nucleus of radon $^{222}_{86}$Rn. An alpha particle (α-particle) and a gamma ray (γ-ray) photon are emitted.
   (i) Write down a nuclear equation to represent this decay.
   (ii) State the conservation laws involved in this nuclear decay.   (6 marks)

(c) The following data are provided for this decay:
mass of $^{226}_{88}$Ra nucleus $= 3.753 \times 10^{-25}$ kg
mass of $^{222}_{86}$Rn nucleus $= 3.686 \times 10^{-25}$ kg
mass of alpha particle $= 0.066 \times 10^{-25}$ kg
speed of light in free space $= 3.00 \times 10^8$ m s^{-1}
the Planck constant $= 6.63 \times 10^{-34}$ J s

For the decay outlined in **(b)**, determine:
   (i) the energy released in the decay
   (ii) the wavelength of the γ-ray photon, given that 4.0% of the energy released becomes γ-ray energy
   (iii) the speed of the α-particle, assuming that the α-particle takes away all the remaining energy   (9 marks)

(d) Use momentum considerations to suggest why the radon nucleus has very much less kinetic energy than the α-particle after the decay.   (3 marks)

Total: 21 marks

Answer to Question 15.1: candidate A

(a) An unstable nucleus emits a helium nucleus to become a more stable different nucleus. An α-particle consists of two protons and two neutrons.

> 🄔 3 marks. A good answer. The α-particle has been identified and its emission from an *unstable* nucleus has been made clear.

(b) (i) $^{226}_{88}$Ra $\rightarrow$ $^{222}_{86}$Rn $+ \frac{4}{2}$He $+ \gamma +$ energy
   (ii) There are several conservation laws governing this reaction. These are:
      1 Conservation of energy
      2 Conservation of momentum
      3 Conservation of proton number
      4 Conservation of nucleon number

e 5 marks. The equation is correct. The candidate should have referred to the conservation of mass/energy, rather than just energy.

(c) (i) Total mass after decay $= 3.752 \times 10^{-25}$ kg
mass loss $= 1.00 \times 10^{-28}$ kg
energy $= mc^2$
$= 1.00 \times 10^{-28} \times 9 \times 10^{16}$
$= 9.0 \times 10^{-12}$ J

(ii) Energy of photon $= \dfrac{hc}{\lambda} = 9.0 \times 10^{-12} \times 0.04 = 3.6 \times 10^{-13}$ J

$6.63 \times 10^{-34} \times 3.00 \times 10^8 = 3.6 \times 10^{-13} \times \lambda$
$\lambda = 5.53 \times 10^{-13}$ J

(iii) Energy of α-particle $= \frac{1}{2} mv^2 = 9.0 \times 10^{-12} \times 0.96 = 8.64 \times 10^{-12}$ J
$\frac{1}{2} \times 0.066 \times 10^{-25} \times v^2 = 8.64 \times 10^{-12}$
$v = 5.12 \times 10^7$ m s^{-1}

e 8 marks. These are straightforward calculations. Do check the answers to make sure they are reasonable. It is easy to make a mistake with powers of ten when using a calculator. In this case, the unit of wavelength has been given as J, rather than m! One mark has been lost needlessly.

(d) The momentum of the α-particle will equal the momentum of the radon nucleus. Since the α-particle has much less mass, its speed will be much greater and so its kinetic energy will be much larger.

e 1 mark. The reasoning as to why the speed of the α-particle is greater is acceptable. However, mass has been ignored when considering kinetic energy! Remember that values have been given, so try to make your discussion quantitative. For example, the speed of the α-particle will be $\frac{222}{4}$ times greater than that of the radon nucleus. Therefore its kinetic energy will be $\left(\frac{4}{222}\right) \times \left(\frac{222}{4}\right)^2$ times greater. For a full discussion, the momentum of the γ-ray photon should be considered. However, it is sufficient here to consider just the radon nucleus and the α-particle.

■ ■ ■

Answer to Question 15.1: candidate B

(a) A nucleus gives off a helium nucleus.

e 1 mark for identifying an α-particle as a helium nucleus. No mention has been made of the parent nucleus being unstable.

(b) (i) $^{226}_{88}\text{Ra} \rightarrow {}^{222}_{86}\text{Rn} + {}^{4}_{2}\text{He} + \gamma\text{-ray}$
(ii) The conservation laws are
1 Conservation of energy
2 Conservation of proton number
3 Conservation of nucleon number

(c) (i) Mass after decay $= 3.752 \times 10^{-25}\,\text{kg}$

mass defect $= 1.00 \times 10^{-28}\,\text{kg}$

energy $= 1.00 \times 10^{-28} \times 9 \times 10^{16}$

$\qquad = 9.0 \times 10^{-12}\,\text{J}$

(ii) Energy of photon $= hf = 9.0 \times 10^{-12} \times \dfrac{4}{100}$

$\qquad\qquad\qquad\quad = 3.6 \times 10^{-13}\,\text{J}$

$f = \dfrac{3.6 \times 10^{-13}}{6.63 \times 10^{-34}}$

$\quad = 5.43 \times 10^{20}$

$\lambda = \dfrac{c}{f} = \dfrac{5.43 \times 10^{20}}{3 \times 10^{8}}$

$\lambda = 1.81 \times 10^{12}\,\text{m}$

(iii) Energy of α-particle $= \frac{1}{2}mv^2 = 9.0 \times 10^{-12} \times \dfrac{96}{100} = 8.64 \times 10^{-12}\,\text{J}$

$\frac{1}{2} \times 0.066 \times 10^{-25} \times v^2 = 8.64 \times 10^{-12}$

$v = 2.6 \times 10^{15}\,\text{m s}^{-1}$

(d) The kinetic energy will be greater because the speed of the α-particle will be bigger.

Question 15.2

(a) Explain what is meant by the terms *random* and *spontaneous* when referring to radioactive decay. (3 marks)

(b) A student obtained the following results for the variation with time of the corrected count rate from a radioactive gas.

Time/s	Corrected count rate/s^{-1}
0	860
30	590
60	400
90	280
120	190
150	130

(i) Explain what is meant by the *corrected* count rate.
(ii) Without plotting a graph, show that the count rate decreases exponentially.
(iii) Using your results in (ii), determine a value for the half-life of the
radioactive gas.

(8 marks)

Total: 11 marks

Answer to Question 15.2: candidate A

(a) Random means that the decay of a nucleus cannot be predicted. There is only a chance
of decay. Spontaneous means that it will occur without warning.

> e 1 mark. The candidate has the idea of randomness but should have said that a
> nucleus has a *constant probability of decay per unit time*. Spontaneity means that
> the probability of decay is unaffected by environmental factors (e.g. temperature).

(b) (i) The count rate from the source after the background count rate has been taken
away.

(ii) If this is true, then $C = C_0 e^{-\lambda t}$

$$\frac{C_0}{C} = e^{\lambda t}$$

$$\ln\left(\frac{C_0}{C}\right) = \lambda t$$

So, $\ln\left(\frac{C_0}{C}\right)$ divided by t should be a constant equal to λ.

Checking $\ln\left(\frac{C_0}{C}\right)$

t	$\ln\left(\frac{C_0}{C}\right)$	$\ln\left(\frac{C_0}{C}\right) \div t$
30	0.377	0.0126
60	0.765	0.0128
90	1.120	0.0124
120	1.510	0.0126
150	1.889	0.0126

Within experimental errors $\ln\left(\frac{C_0}{C}\right) \div t$ is constant so the count rate drops exponentially.

(iii) Average value of $\ln\left(\frac{C_0}{C}\right) \div t$ is $0.0126 = \lambda$

$$\lambda t_{1/2} = \ln 2$$

$$t_{1/2} = \frac{0.693}{0.0126} = 55 \text{ s}$$

> e 8 marks. The explanation in part (i) of *corrected count rate* is satisfactory. In part (ii)
> the explanation as to why $\ln\left(\frac{C_0}{C}\right) \div t$ is calculated is good. In part (iii) an average
> value was used for the calculation of the half-life, rather than taking one value; this
> is always good technique.

Answer to Question 15.2: candidate B

(a) Random and spontaneous mean the nucleus has a chance of decaying.

> e No marks. It is a common misconception that random is an alternative word for spontaneous. This is not true. Remember to state the 'probability (chance) per unit time' rather than just 'chance'.

(b) (i) The count rate after allowance has been made for background.

(ii) $C = C_0 e^{-\lambda t}$

taking logs to base e, $\ln C = \ln C_0 - \lambda t$

If it is true, $\dfrac{\ln C}{t}$ should be constant and equal to λ.

t	$\ln C$	$\dfrac{\ln C}{t}$
30	6.380	0.2127
60	5.991	0.0999
90	5.635	0.0626
120	5.247	0.0437
150	4.868	0.0325

Within experimental errors $\dfrac{\ln C}{t}$ is constant, apart from $t = 30\,\text{s}$, so the count rate drops exponentially.

(iii) Ignoring $t = 30\,\text{s}$, average value of $\dfrac{\ln C}{t}$ is $0.0597 = \lambda$

$\lambda t_{1/2} = \ln 2$

$t_{1/2} = \dfrac{0.693}{0.0597} = 11.6\,\text{s}$

> e 4 marks. No marks are scored in part (i) because the candidate has not said how the allowance is made. In part (ii) a correct equation is given but the subsequent analysis is incorrect. Assuming that the candidate had calculated values for λ in part (ii), then the working in part (iii) would have been correct and two marks are therefore scored.

Question 15.3

(a) Explain what is meant by a fission reaction. (3 marks)

(b) The following equation is a possible nuclear reaction for uranium-235:

$$^{235}_{92}\text{U} + ^{1}_{0}\text{n} \rightarrow ^{141}_{56}\text{Ba} + ^{92}_{36}\text{Kr} + 3^{1}_{0}\text{n} + \text{energy}$$

(i) Explain how this reaction could lead to a chain reaction.
(ii) The graph below shows the variation with nucleon number of the binding energy per nucleon for some nuclei.

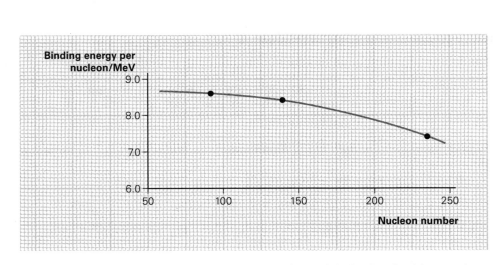

Use the graph to determine the energy released, in joules, in this reaction. The elementary charge = 1.6×10^{-19} C.

(iii) State the forms in which this released energy would be seen.   (9 marks)

Total: 12 marks

Answer to Question 15.3: candidate A

(a) A nucleus splits into two parts with approximately equal masses, together with some neutrons and energy.

> e 2 marks. The candidate has failed to mention that the parent nucleus must be massive.

(b) (i) If there are many uranium nuclei, then the three neutrons will go on to cause three more nuclei to react. These three nuclei will release nine neutrons which can go on to other nuclei etc.

$1 \rightarrow 3 \rightarrow 9 \rightarrow 27 \rightarrow 81 \rightarrow 243 \rightarrow 7129$ etc.

This is a chain reaction because the reaction snowballs.

(ii) Change in binding energy $= (92 \times 8.6) + (141 \times 8.42) - (235 \times 7.4)$
$$= 239 \,\text{MeV}$$
$$= 3.8 \times 10^{-17} \,\text{J}$$

(iii) The γ-ray has some energy but most is seen as kinetic energy of the daughter nuclei and the neutrons.

> e 8 marks. The chain reaction is described satisfactorily. In part (ii) the only error is the conversion of MeV to joules (1 MeV = 1.6×10^{-13} J). In part (iii) it would have been better to refer to a γ-ray photon, rather than to just a γ-ray.

Answer to Question 15.3: candidate B

(a) A large nucleus splits into two, giving off some neutrons.

> 🄴 1 mark. The candidate realises that fission occurs only with massive nuclei. The sizes of the fission fragments and the release of energy have not been included.

(b) (i) The three neutrons can go on to fission more nuclei. The neutrons released will then fission even more nuclei, causing an avalanche effect.

(ii) Energy released $= 8.6 + 8.4 - 7.4$

$$= 9.6\,\text{MeV}$$
$$= 1.5 \times 10^{-12}\,\text{J}$$

(iii) The energy is seen as heat.

> 🄴 2 marks. In part (i) the candidate has the idea of what is meant by a chain reaction, but that idea is very poorly expressed. This is a good example of an answer where a sketch diagram would have been very useful. In part (ii) no account has been taken of the nucleon numbers of the nuclei involved. This is a common mistake. The conversion to joules of MeV is correct. The answer in part (iii) is not worthy of AS/A-level and would be given no marks. Candidates should know that any rise in temperature is as a result of an increase in kinetic energy of the fission fragments.

Question 15.4

(a) (i) Summarise the experimental observations made in the alpha-particle (α-particle) scattering experiment.

(ii) Explain how these observations provided evidence for a nuclear model of the atom.

(iii) Suggest why, in order to investigate the composition of the nucleus, particles other than alpha particles are necessary. (8 marks)

(b) Some particles are listed in the table below.

Particle	Gauge boson	Lepton	Hadron	
			Meson	Baryon
Proton				
Neutron				
Electron				
Photon				
Neutrino				

Copy the table and complete it by placing a tick in the relevant boxes to indicate the classification of each particle. (3 marks)

Total: 11 marks

Answer to Question 15.4: candidate A

(a) (i) Most α-particles passed through the foil with small deviations. About 1 in 10^4 suffered a deviation greater than 90°.

(ii) Because most passed through, the atom must be almost all empty space. The few bouncing back means that there must be a central core which is small, massive and charged.

(iii) To investigate the nucleus, particles must go into the nucleus. This means they must have large energies. α-particles do not have enough energy.

e 8 marks. In parts (i) and (ii) the observations are summarised and related to the conclusions drawn from them. Although not expressed very well, the candidate appreciates that, in order to investigate the nucleus, the nucleus must be penetrated so that the constituent particles can be identified and investigated. This requires incident particles with much higher energies than α-particles.

(b)

Particle	Gauge boson	Lepton	Hadron	
			Meson	Baryon
Proton				✔
Neutron				✔
Electron		✔		
Photon	✔			
Neutrino		✔		

e 3 marks. All correct.

Answer to Question 15.4: candidate B

(a) (i) Most α-particles passed straight through the foil but a few bounced back from the foil.

(ii) The atom is mostly empty with a very small nucleus containing most of the mass of the atom and is positively charged.

(iii) The particles must have much more energy to smash the nucleus.

e 3 marks. In part (i) the candidate has the correct idea, but the answer is poorly expressed. No mention is made of deviations and their relative magnitudes, or the proportion of α-particles deviated through more than 90°. The wording 'passed straight through' should be avoided because it implies no deviation whatsoever. Part (ii) shows a very common mistake: the nuclear atom has been described satisfactorily for this question but the description has not been linked with the experimental observations. In part (iii) the answer is very superficial and is not worth a mark.

(b)

Particle	Gauge boson	Lepton	Hadron	
			Meson	Baryon
Proton				✔
Neutron				✔
Electron		✔		
Photon			✔	
Neutrino	✔			

ⓔ 1 mark. Only three particles have been classified correctly.

Mass-energy

Question 16.1

(a) Explain what is meant by the *binding energy* of a nucleus. (2 marks)

(b) In the JET nuclear fusion reactor project, a possible nuclear fusion reaction is:

$$^2_1H + ^3_1H \rightarrow ^4_2He + X + 17.6\,MeV$$

(i) Explain what is meant by a *fusion* reaction.
(ii) Give the nuclear representation of the particle X and its identity.
(iii) Convert 17.6 MeV to energy in joules.

The elementary charge is 1.6×10^{-19} C. (6 marks)

(c) One mole of helium has a mass of 4.0×10^{-3} kg. The speed of light in free space is 3.0×10^8 m s^{-1}.
(i) Calculate the energy released when 1.0 kg of helium ^4_2He is produced in the reaction in **(b)**. The Avogadro constant is 6.0×10^{23} mol^{-1}.
(ii) Determine the mass equivalent of the energy in (i). (4 marks)

(d) The energy released when 1.0 kg of coal is burned is 29 MJ. Calculate the mass of coal that would need to be burned to produce the same amount of energy as in **(c)** (i). (2 marks)

Total: 14 marks

■ ■ ■

Answer to Question 16.1: candidate A

(a) Binding energy of a nucleus is the energy required to separate all the nucleons of the nucleus to infinity.

 2 marks. A correct definition. Many candidates fail to say that, finally, the nucleons must be infinitely far apart.

(b) (i) Fusion occurs when two nuclei combine to form a more massive nucleus with the release of energy.
(ii) ^1_0n. It is a neutron.
(iii) 1 eV = 1.6×10^{-19} J
17.6 MeV = $17.6 \times 10^6 \times 1.6 \times 10^{-19} = 2.8 \times 10^{-12}$ J

 5 marks. In part (i) the candidate should have stated that the nuclei are *low mass* nuclei. Parts (ii) and (iii) are answered well.

(c) (i) In 1 kg of helium there are $\frac{1}{4}N_A$ atoms = 1.5×10^{23}
Energy = $1.5 \times 10^{23} \times 2.8 \times 10^{-12} = 4.2 \times 10^{11}$ J

(ii) $E = mc^2$

$4.2 \times 10^{11} = m \times 9 \times 10^{16}$

$m = 4.7 \times 10^{-6}\,\text{kg}$

> ⓔ 3 marks. There is an error in part (i). There are $250N_A$ nuclei in 1 kg of helium. The remainder of the section is correct, when based on this wrong value for the number of nuclei, and would score full marks. The correct answers are $4.2 \times 10^{14}\,\text{J}$ and $4.7 \times 10^{-3}\,\text{kg}$.

(d) Mass $= \dfrac{4.2 \times 10^{11}}{29 \times 10^6} = 1.45 \times 10^4\,\text{kg}$

> ⓔ 2 marks. The calculation is correct when based on the candidate's value for the energy release. A three significant figure answer from two significant figure data would be excused.

■ ■ ■

Answer to Question 16.1: candidate B

(a) It is the energy released when a nucleus is formed from its protons and neutrons.

> ⓔ 1 mark. The candidate has not said that the neutrons and protons should, initially, be infinitely far apart.

(b) (i) Fusion occurs when two nuclei fuse to form a bigger nucleus.

(ii) ^1_0n. It is a neutron.

(iii) $17.6\,\text{MeV} = 17.6 \times 10^6 \times 1.6 \times 10^{-19} = 2.8 \times 10^{-12}\,\text{J}$

> ⓔ 4 marks. No marks have been scored in part (i): the idea that light nuclei form a more massive nucleus has not been included and 'Fusion occurs when two nuclei fuse…' doesn't provide any information about the meaning of fusion. Parts (ii) and (iii) are satisfactory.

(c) (i) There are $6.0 \times 10^{23} \times \dfrac{1}{0.004} = 1.5 \times 10^{26}$ atoms

Energy $= 1.5 \times 10^{26} \times 2.8 \times 10^{-12} = 4.2 \times 10^{14}\,\text{J}$

(ii) $4.2 \times 10^{14} = m \times 9 \times 10^8$

$m = 4.7 \times 10^5\,\text{kg}$

> ⓔ 2 marks. Part (i) is correct. No marks have been scored in part (ii). The candidate has not used c^2 ($= 9 \times 10^{16}$). Is this an error of physics or an arithmetical error? The examiner cannot tell. If the candidate had written down the formula $E = mc^2$, then a mark would have been given. Always write down formulae before substituting. The candidate should have thought about the answer and then realised that it is far too large a mass.

(d) Mass $= \dfrac{4.2 \times 10^{14}}{29 \times 10^6} = 1.4 \times 10^7\,\text{kg}$

> ⓔ 2 marks. The calculation is correct.

Question 16.2

(a) State and explain the conditions required for a nuclear fusion reaction to take place. (6 marks)

(b) The solar constant is a measure of the power received per unit area from the Sun by the Earth. Its value is $1.2\,kW\,m^{-2}$ and the distance from the Sun to the Earth is $1.5 \times 10^8\,km$. Calculate:
(i) the power output of the Sun
(ii) the loss in mass per unit time of the Sun
The speed of light is $3.0 \times 10^8\,m\,s^{-1}$. (6 marks)

(c) The life of the Sun has been estimated to be about 10^{10} years and its mass is $2 \times 10^{30}\,kg$. Comment on this value for the lifetime with reference to your answer in (b) (ii). (3 marks)

Total: 15 marks

■ ■ ■

Answer to Question 16.2: candidate A

(a) The temperature must be very high so that the nuclei are moving fast enough to hit each other. Also the pressure must be high so that there is a good chance that nuclei will be near each other.

> e 4 marks for two conditions with adequate explanation. The candidate has not mentioned that the nuclei must be of low mass (i.e. well below ^{56}Fe) so that the resultant nucleus, which is more massive, has a higher binding energy per nucleon.

(b) (i) Area of sphere $= 4\pi r^2 = 4 \times \pi \times (1.5 \times 10^{11})^2$
$$= 2.83 \times 10^{23}\,m^2$$
Power of Sun $= 2.83 \times 10^{23} \times 1.2 = 3.4 \times 10^{23}\,kW$
(ii) $E = mc^2$
$3.4 \times 10^{26} = m \times 9 \times 10^{16}$
Mass lost per second $= 3.8 \times 10^9\,kg$

> e 6 marks. The calculations are correct, with adequate explanation. It would have been better to give the answer to part (i) as $3.4 \times 10^{26}\,W$, but the answer given is acceptable and it was converted to watts before substitution in part (ii).

(c) Mass lost in lifetime $= 3.8 \times 10^9 \times 10^{10} \times 365 \times 24 \times 3600$
$$= 1.2 \times 10^{27}\,kg$$
This is less than one-thousandth of the total mass. However, if a lot of the mass is lost, then the temperature and pressure would be too small for fusion to take place.

> e 3 marks. The numerical part is correct. The candidate is not expected to have an understanding of astrophysics. The suggestion is sensible and is based on the answer to part (a).

■ ■ ■

Answer to Question 16.2: candidate B

(a) The temperature and the pressure must be very high so that the nuclei have sufficient energy.

> 🄴 2 marks. Two conditions have been given but the explanation is inadequate. The candidate has not mentioned that the nuclei must be of low mass.

(b) (i) Area $= 4 \times \pi \times (1.5 \times 10^{11})^2 = 2.83 \times 10^{23}\, m^2$
Power of Sun $= 2.83 \times 10^{23} \times 1.2 = 3.4 \times 10^{23}\, kW$

(ii) $E = mc^2$
$3.4 \times 10^{23} = m \times 9 \times 10^{16}$
mass lost per second $= 3.8 \times 10^6\, kg$

> 🄴 5 marks. The calculations in part (i) are correct, with adequate explanation. The answer was given in kW but this was not converted to watts before substitution in part (ii), resulting in the loss of a mark.

(c) Mass lost in lifetime $= 3.8 \times 10^6 \times 10^{10} \times 365 \times 24 \times 60$
$$= 2 \times 10^{22}\, kg$$
This is very small compared to the total mass. Mass must be lost in other ways.

> 🄴 1 mark. A numerical error has been made in the conversion of years to seconds. This type of error is quite common. The suggestion has not been given any marks because it implies that the total mass of the Sun will be lost in its lifetime. This would contradict the answer in (a).

17

Astrophysics and cosmology

Question 17.1

(a) Suggest values for:
  (i) the distance, in kilometres, of the Sun from the Earth
  (ii) the mass, in kilograms, of the Sun
  (iii) one light-year (ly) in kilometres
  (iv) one parsec (pc) in light-years (4 marks)

(b) Draw sketch diagrams to show the shape of the Milky Way galaxy. On your diagrams give relevant distances and mark with the letter S the position of the Sun. (5 marks)

(c) The Sun takes approximately 200 million years to make one journey round the galaxy. Estimate the linear speed of the Sun, in $km\,h^{-1}$, in its orbit. (3 marks)

Total: 12 marks

■ ■ ■

Answer to Question 17.1: candidate A

(a) (i) $1.5 \times 10^8\,km$
  (ii) $10^{30}\,kg$
  (iii) $365 \times 24 \times 3600 \times 3 \times 10^5 = 9.46 \times 10^{12}\,km$
  (iv) $3.26\,ly$

  🄔 4 marks. The values expected are, in fact, only estimates. These values are well within tolerance. The mass of the Sun is generally taken to be about $2 \times 10^{30}\,kg$.

(b)

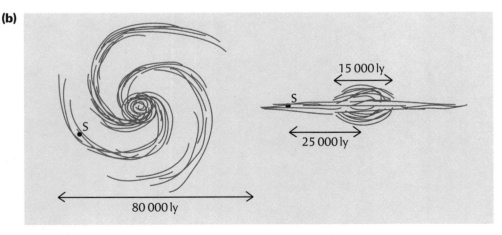

  🄔 5 marks. A spiral galaxy is not an easy object to draw! The candidate shows clearly the spiral nature together with the central bulge. The position of the Sun is satisfactory (about two-thirds of 'the way out' on one of the spiral arms. Some dimensions have been given. Others could have included the thickness of the disc, but with this mark allocation not all distances would be expected.

(c) Speed $= \dfrac{2\pi r}{T}$

$T = 200 \times 10^6 \times 365 \times 24 = 1.75 \times 10^{12}$ hours

$r = 25\,000 \times 9.46 \times 10^{12} = 2.37 \times 10^{17}$ km

speed $= \dfrac{2\pi \times 2.37 \times 10^{17}}{} = 8.5 \times 10^5\,\text{km h}^{-1}$

> 3 marks. The candidate has carefully worked through each stage of the calculation. The work may seem to be rather pedestrian, but it is advisable in calculations such as this because it is so easy to make a mistake with the calculator. Showing the separate stages allows the examiner to see where any error has occurred and to give marks for all correct work.

■ ■ ■

Answer to Question 17.1: candidate B

(a) (i) 1.5×10^8 km

(ii) 10^{25} kg

(iii) 9.5×10^{12} km

(iv) 3.3 ly

> 3 marks. In part (ii) it is likely that the candidate has confused the mass of the Earth with that of the Sun.

(b)

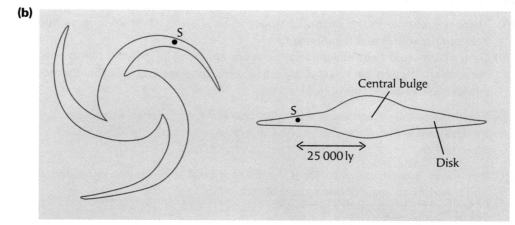

> 3 marks. Although the diagram is of poor quality, the candidate does indicate the spiral nature together with the central bulge. The position of the Sun is satisfactory. Some further dimensions should have been included.

(c) Speed $= \dfrac{2\pi \times 25\,000 \times 9.5 \times 10^{12}}{200 \times 10^6 \times 365 \times 24}$

$= 2.7 \times 10^5 \, \text{km h}^{-1}$

> 2 marks. The substitution appears to be correct. However, there is an error in working out the final answer. Note that, if the working had not been shown, then the wrong answer on its own would have scored no marks.

Question 17.2

(a) Explain briefly:
  (i) what is meant by the *Doppler effect*
  (ii) how observations of the Doppler effect lead to Hubble's law   (5 marks)

(b) One wavelength in the line spectrum of calcium when measured for a source in the laboratory is found to be 393.3 nm. The same spectral line has a wavelength of 401.8 nm when measured in the light from a distant galaxy. The Hubble constant is 80 km s^{-1} Mpc^{-1}.
  (i)   State and explain the unit Mpc.
  (ii) Calculate the velocity of the galaxy relative to Earth. The speed of light is $3.0 \times 10^8 \, \text{m s}^{-1}$.
  (iii) Estimate how long it would take for light to travel from the galaxy to Earth.
                                                          (9 marks)
                                                      Total 14 marks

Answer to Question 17.2: candidate A

(a) (i) There is a change in the measured frequency of a wave when there is motion between the source and the observer.
  (ii) Galaxies are found to be moving away from one another. The speed v at which they move apart is given by $v = H_0 d$. H_0 is Hubble's constant and d is the distance between the galaxies. This is Hubble's law.

> 5 marks. In part (i) the candidate has made it clear that it is the observed frequency that changes. In part (ii) the discussion has not been restricted to movement away from the Earth.

(b) (i) It is one million parsecs. A parsec is the distance away at which one astronomical unit subtends an angle of one second of arc.

  (ii) For the Doppler effect $\dfrac{\Delta\lambda}{\lambda} = \dfrac{v}{c}$

$\Delta\lambda = 8.5 \, \text{nm}$

$\dfrac{8.5}{393.3} = \dfrac{v}{3.0 \times 10^8}$

$v = 6.5 \times 10^6 \, \text{m s}^{-1}$

(iii) Using the Hubble law
$6.5 \times 10^6 = 80 \times d$
$d = 81\,250\,\text{Mpc}$
$1\,\text{pc} = 3.26\,\text{ly}$
$d = 2.65 \times 10^5$ million light-years
Time taken $= 2.65 \times 10^{11}$ years

> 7 marks. The explanation in part (i) is satisfactory. Some candidates like to draw a sketch to assist with the explanation. In part (ii) a common error has been made: candidates were asked to find the velocity and so the candidate should have said that this is the speed away from Earth. The explanation in part (iii) is good and it is clear where the one mistake has been made. The Hubble constant should have been converted to $\text{m s}^{-1}\,\text{Mpc}^{-1}$. The correct answer is 265 million years.

■ ■ ■

Answer to Question 17.2: candidate B

(a) (i) The frequency of light changes when the source moves.
(ii) The speed v at which a galaxy is moving away from the Earth is equal to $H_0 d$. d is the distance from Earth.

> 2 marks. Part (i) is poorly answered. The concept that it is the *observed* frequency that changes as a result of *relative* motion is not made clear. In part (ii) the discussion has been restricted to movement away from the Earth and H_0 has not been identified as being the Hubble constant.

(b) (i) It is a megaparsec. One parsec is the distance for the Earth–Sun distance to give one second of arc.

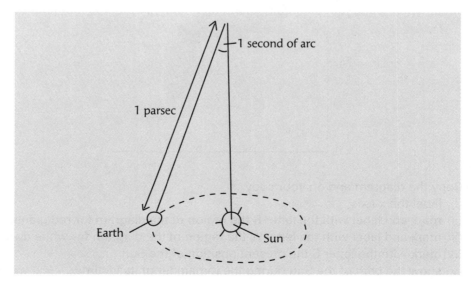

(ii) $\dfrac{\Delta\lambda}{\lambda} = \dfrac{v}{c}$

$\dfrac{8.5}{401.8} = \dfrac{v}{3.0 \times 10^8}$

$v = 6.35 \times 10^6\,\text{m s}^{-1}$

(iii) Using the Hubble law

$6.35 \times 10^3 = 80 \times d$

$d = 79\,\text{Mpc}$

$1\,\text{pc} = 5\,\text{ly}$

Time taken = 395 million years

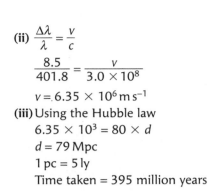 5 marks. The explanation in part (i) is poor but the diagram provides adequate compensation. However, 1 mark has been lost because 'mega' has not been explained. In part (ii) the observed frequency, rather than the true frequency, was substituted into the formula for the Doppler effect. Also, it was not stated that the galaxy is receding. The explanation in part (iii) is satisfactory but the conversion from parsec to light-year is erroneous.

Question 17.3

(a) Below is an incomplete Hertzsprung–Russell (H–R) diagram.

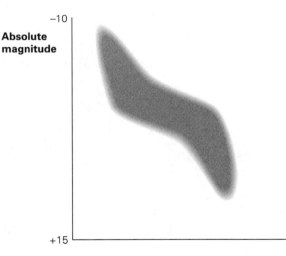

Copy the diagram and on your copy:

(i) label the x-axis

(ii) mark and label with the letter R the region of the diagram for red giants

(iii) mark and label with the letter W the region of the diagram for white dwarfs

(iv) mark with the letter S the present position of the Sun

(v) show the path of the Sun during the remainder of its lifetime   (7 marks)

(b) The *y*-axis of the diagram is labelled 'absolute magnitude'.
  (i)   Explain what is meant by *absolute magnitude*.
  (ii) Two stars A and B have absolute magnitudes of +1.0 and +6.0 respectively.
        The ratio of the intensities of the light emitted by these stars is 100.
        (1) State which star is the brighter.
        (2) State and explain the ratio of the intensities of light from two stars that
            have a difference in absolute magnitude of 1.0.   (6 marks)

Total: 13 marks

Answer to Question 17.3: candidate A

(a)

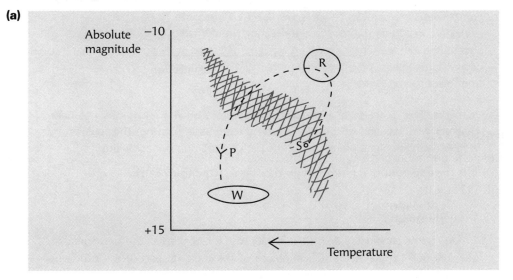

🖉 6 marks. Note that the direction of increasing temperature on the *x*-axis has been
   marked. However, there is no indication of the magnitude of the temperatures
   (2500 K to 40 000 K). The remainder of the diagram is acceptable. Note that regions
   cannot be given precise locations on the diagram.

(b) (i) It is the brightness of a star if it is 10 pc from Earth.
   (ii) (1) Star A.
        (2) Each equal change in magnitude is the same change in ratio of intensity.
            $R^6 = 100$
            $R = 2.2$
            (The answer should be 2.5.)

🖉 5 marks. Although the candidate realises that there is an error in part (ii) (2), the error
   has not been spotted. The correct working is $R^{6-1} = 100$.

Answer to Question 17.3: candidate B

(a)

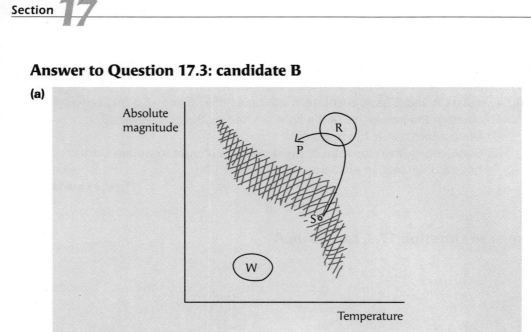

> 4 marks. The direction of increasing temperature on the *x*-axis has not been marked. An indication of the relevant temperatures is missing. The regions are located with adequate precision but the path for the Sun is incomplete.

(b) (i) It is the apparent magnitude when the star is 10 pc from Earth.

(ii) (1) Star B

(2) Ratio is $\dfrac{100}{5} = 20$

> 1 mark. Some candidates do have difficulty with the concept of stellar magnitude, possibly due to the mathematical nature of the topic. In part (i) the candidate should have explained that apparent magnitude is brightness. Part (ii) is incorrect — the lower the number, the greater the brightness. There is very little explanation in part (ii) (2), and the indication is that the candidate does not appreciate the concept involved.

Medical physics

Question 18.1

(a) (i) With reference to the eye, explain what is meant by *accommodation*.
(ii) Explain briefly how accommodation is achieved in the human
eye. (5 marks)

(b) An elderly person cannot see clearly objects that are closer than 100 cm to
his eyes. Determine the lenses he should wear in order to be able to see
objects placed at the normal near point of the eye. (4 marks)

(c) (i) The diameter of the pupil of a particular eye can vary between 2.0 mm and
6.0 mm. Assuming constant intensity of light incident on the eye, calculate
the ratio of the light power on the retina for the maximum and minimum
diameters of the pupil.
(ii) It is known that the ratio of the maximum sensitivity to the minimum
sensitivity of the eye is greater than 10^6. Comment on your answer to (i)
with reference to this range of sensitivity. (4 marks)

Total: 13 marks

■ ■ ■

Answer to Question 18.1: candidate A

(a) (i) Accommodation is the means by which the eye can focus on objects at different
distances from the eye.
(ii) The ciliary muscles change the shape of the lens and also its focal length.

> 🄴 4 marks. Part (i) has been answered well. In part (ii) the candidate should have
> mentioned that the fatter the lens, the nearer the object can be to the eye for a
> focused image.

(b) Near point is about 25 cm from the eye.

Using formula $\dfrac{1}{u} + \dfrac{1}{v} = \dfrac{1}{f}$

$\dfrac{1}{25} - \dfrac{1}{100} = \dfrac{1}{f}$

focal length = 33.3 cm and the lens is convex.

> 🄴 4 marks. A well-explained calculation. Note that the type of lens has been
> stated, along with its focal length. The question did not ask for the power of the
> lens, but an equally acceptable final answer would have been 'convex lens of
> power 3.0 D'.

(c) (i) Ratio is the ratio of the areas

area $= \frac{1}{4}\pi d^2 \propto d^2$

ratio is $\frac{6^2}{2^2} = 9.0$

(ii) The iris can only adjust for light intensity changes of 9. The retina cannot have a constant sensitivity.

> 🄴 3 marks. The calculation in part (i) is correct. Some further detail was expected in part (ii). For example, the candidate has not mentioned that the adjustment of the iris would be to attempt to maintain constant power on the retina. Alternatively, some reference to rods and cones would have been appropriate.

■ ■ ■

Answer to Question 18.1: candidate B

(a) (i) It allows the eye to focus on objects at different distances away.
(ii) The shape of the lens is changed by muscles.

> 🄴 2 marks. The answer to part (i) is barely satisfactory. In part (ii) the candidate has not given the name of the muscles or discussed the effect of the shape of the lens on focusing. With 5 marks available, the candidate should have realised that more detail would be required.

(b) $\frac{1}{100} - \frac{1}{25} = \frac{1}{f}$

focal length $= -33.3\,\text{cm}$

> 🄴 1 mark. The candidate has not written down the formula before attempting the substitution. The object and image have been confused as regards which is virtual. The type of lens has not been stated.

(c) (i) Ratio is $\frac{6}{2} = 3$
(ii) The retina has a variable sensitivity.

> 🄴 1 mark. The calculation in part (i) scores no marks since area is not involved. The candidate should have realised that, with four marks available for this section of the question, some further detail or explanation is required. The candidate should have realised that the question involves long-sight, where a convex lens is required.

Question 18.2

(a) Describe briefly how sound vibrations are transmitted through the ear.

(8 marks)

(b) The diagram below shows the variation with frequency of the minimum intensity of sound that can be detected by a person with normal hearing.

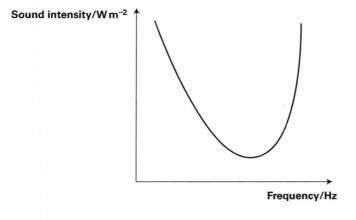

Sound intensity/W m^{-2}

Frequency/Hz

Copy the diagram and mark any significant values on the axes. (4 marks)

(c) The average intensity of the noise in a workplace is found to be 6.3 mW m^{-2}.
  (i) Calculate the intensity level of this noise.
  (ii) Comment on your answer in (i). (3 marks)

Total: 15 marks

■ ■ ■

Answer to Question 18.2: candidate A

(a) The sound waves are collected by the pinna and travel down the auditory canal to the ear drum. This vibrates, making bones in the middle ear vibrate.

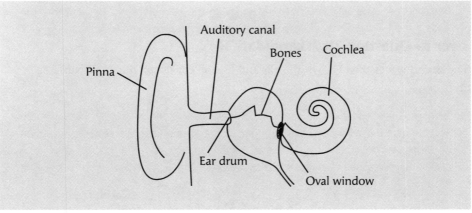

The bones transfer the vibrations to the oval window and this makes hairs in the cochlea vibrate. Nerves attached to the hairs transfer electrical signals to the brain.

e 6 marks. It is always advisable to draw a diagram when giving a description such as this. The basic outline is satisfactory but some further detail is required. For example, the function of the malleus, incus and stapes (a lever system to increase the pressure changes on the oval window). Also, the fibres of the basilar membrane vary in length and stiffness so as to respond to different frequencies.

(b)

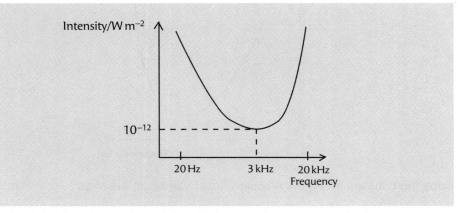

    4 marks. The frequency range shown is within normal limits and the minimum intensity at 3 kHz has been given.

(c) (i) Intensity level $= 10\log\left(\dfrac{I}{10^{-12}}\right)$

$\qquad\qquad\qquad = 10\log\left(6.3 \times 10^{9}\right)$

$\qquad\qquad\qquad = 98\,\text{dB}$

(ii) This intensity level could cause deafness and must be reduced.

    3 marks. The calculation is set out well and a sensible comment has been made.

■ ■ ■

Answer to Question 18.2: candidate B

(a) The sound waves travel down the ear and hit the ear drum, making it vibrate.

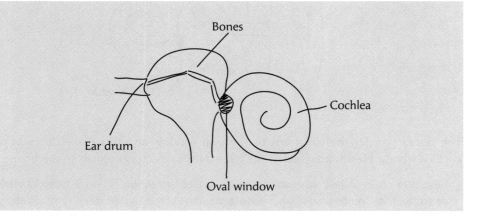

Bones make the oval window vibrate. Hairs in the cochlea vibrate, sending signals to the brain.

> 3 marks. The description is incomplete and lacks detail. A diagram has been drawn but it is of limited value. If a diagram is drawn, then it must be made worthwhile. The candidate should appreciate that some comment is required for each stage of the transmission through the ear — outer ear, ear drum, middle ear, oval window, cochlea and nerves.

(b)

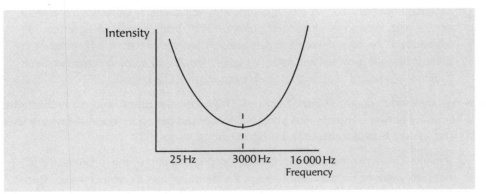

> 3 marks. Relevant frequencies are shown but the intensity at the threshold of hearing has not been included.

(c) (i) Intensity level $= 10\log\left(\dfrac{I}{I_0}\right)$

$$= 10\log(6.3 \times 10^6)$$
$$= 68\,\text{dB}$$

(ii) This intensity level would be allowed because it would not cause deafness.

> 2 marks. The formula for the calculation has been given but the substitution for I_0 is incorrect. The subsequent calculation is correct and is given one mark. The comment has been given credit because it is relevant to the candidate's value for the intensity level.

Question 18.3

(a) Outline briefly the process by which ultrasound is generated and used for diagnostic purposes.

(8 marks)

(b) Suggest two different situations in medicine where X-rays are used in preference to ultrasound. For each suggestion, state why this choice is made.

(4 marks)

Total: 12 marks

Answer to Question 18.3: candidate A

(a) A high-frequency voltage pulse is applied across a piezo-electric crystal. The crystal produces a pulse of ultrasound. This pulse travels into the body and is reflected at a boundary between two types of tissue. The reflected wave is picked up by the crystal. It produces an electrical signal which is amplified and fed into a computer. The output is displayed on a screen.

> e 5 marks. The basic outline is satisfactory but there are two omissions. The candidate has not said that the high-frequency voltage pulse causes the crystal to vibrate and, when the ultrasound pulse returns to the crystal, it again causes vibrations of the crystal. The candidate should have stated that, at a boundary, the pulse is partially reflected and partially transmitted. A very common error has been avoided — a failure to state that the ultrasound must be pulsed.

(b) X-rays are used to diagnose bone fractures. This is because ultrasound is all reflected at a boundary between muscle and bone but X-rays can penetrate bone. X-rays are used to treat cancer because ultrasound is harmless.

> e 3 marks. One mark has been lost because stating that ultrasound is harmless does not give a reason for the use of X-rays. Furthermore, it is a common misconception that ultrasound is harmless — it can cause damage to cells.

■ ■ ■

Answer to Question 18.3: candidate B

(a) Ultrasound is produced when a high-frequency voltage is applied to a crystal. The ultrasound is reflected off tissues in the body and is picked up by the crystal. The signals produced in the crystal are fed to a computer and displayed on a screen.

> e 2 marks. Several important facts are missing. There is no mention that the alternating voltage causes the *piezo-electric* crystal to vibrate and that ultrasound *pulses* must be generated. It is not clear that the ultrasound reflections occur at *boundaries* between two types of tissue. Also, it should be stated that the *electrical* signals generated in the crystal must be *amplified*.

(b) X-rays are used to look at broken bones because they are more penetrating. Also, if a barium meal is given to a patient, they can be used to detect stomach ulcers.

> e 1 mark only. The two suggestions are, in fact, both diagnostic and so are treated as one example. A mark has not been awarded for the reason because it is vague. Both X-rays and ultrasound can penetrate soft tissue.

R51999